RESEARCH IN MARITIME HISTORY
NO. 54

FISHERMEN, THE FISHING INDUSTRY AND THE GREAT WAR AT SEA

A FORGOTTEN HISTORY?

ROBB ROBINSON

MAILING ADDRESS

Liverpool University Press
4 Cambridge Street
Liverpool L69 7ZU
United Kingdom

Recently Published Books in the Series

46: Lewis R. Fischer and Even Lange (eds), *New Directions in Norwegian Maritime History*

47: John Armstrong and David M. Williams, *The Impact of Technological Change: The Early Steamship in Britain*

48: Ralph Davis, *The Rise of the English Shipping Industry in the Seventeenth and Eighteenth Centuries*

49: Gordon Boyce, *The Growth and Dissolution of a Large-Scale Business Enterprise: The Furness Interest 1892–1919*

50: Torsten Feys, *The Battle for the Migrants: The Introduction of Steamshipping on the North Atlantic and Its Impact on the European Exodus*

51: Anthony Slaven and Hugh Murphy (eds), *Crossing the Bar: An Oral History of the British Shipbuilding, Ship Repairing and Marine Engine-Building Industries in the Age of Decline, 1956–1990*

52: Olaf U. Janzen, *War and Trade in Eighteenth-Century Newfoundland*

53: Christopher W. Miller, *Planning and Profits: British Naval Armaments Manufacture and the Military-Industrial Complex, 1918–1941*

RESEARCH IN MARITIME HISTORY
NO. 54

Fishermen, the Fishing Industry and the Great War at Sea

A Forgotten History?

ROBB ROBINSON

First published 2019 by
Liverpool University Press
4 Cambridge Street
Liverpool
L69 7ZU

British Library Cataloguing-in-Publication data
A British Library CIP record is available

ISBN 978-1-78694-175-6

Typeset by Carnegie Book Production, Lancaster
Printed and bound by CPI Group (UK) Ltd, Croydon, CR0 4YY

Contents

Tables

Photographs

Figures

Introduction

On 21 October 1904, on the eve of the 99th anniversary of the Battle of Trafalgar, warships from the Baltic fleet of the Imperial Russian Navy, under the command of Admiral Rozhestvensky, entered the North Sea. They were on a war footing, and later, in the evening of the same day, the Russians bore down on the trawlers of Hull's Gamecock fishing fleet. Powerful searchlights illuminated the seas and the warships opened fire. For 20 minutes, the unarmed fishing vessels were subjected to an unrelenting bombardment.[1] The steam trawler *Crane* received the heaviest hits: skipper George Smith and mate John Leggott were decapitated by the shells which smashed into their vessel. The *Crane* eventually sank, and the rest of the crew were rescued by the trawler *Gull*, but two other fishing vessels, *Mouleim* and *Mino*, which were also subject to heavy fire, were badly damaged yet stayed afloat. The order to cease fire came just a few minutes before 1 a.m. and the Russian warships hurried off into the night towards the English Channel without pausing to inspect the carnage they had created. Meanwhile, the stunned crews of the Gamecock trawler fleet gave up fishing, stowed away their gear, and steamed off in great haste, heading for the Humber and their home port of Hull.[2]

The Dogger Bank incident, or 'Russian Outrage', as it became known, was a tragic mistake. Russia was not at war with Britain but with Japan. Russia's Baltic fleet, consisting of 42 warships of varying age and size, was on the first leg of what was to be an 18,000 mile voyage to the Far East, the intention being to reinforce the navy vessels already there that had endured a series of serious setbacks in their nation's vicious conflict with Japan. To that date the Japanese had surprised and outwitted the Russians in many actions and Rozhestvensky and his command – expecting attack at almost any time on their voyage, despite the great distances involved – had mistaken these civilian trawlers for Japanese torpedo boats.[3]

The repercussions of the attack were immediate and global, the British media were incensed; there was talk of declaring war on Russia and of the

1 Robb Robinson, *Trawling: The Rise and Fall of the British Trawl Fishery* (Exeter, 1886), 125–126.

2 *Eastern Morning News*, 23 October 1904.

3 Robinson, *Trawling*, 125.

Royal Navy attacking this fleet, but in the event Rozhestvensky and his ironclad armada were left to continue their convoluted cruise to the Far East. Perhaps not surprisingly, most of these Russian warships were subsequently destroyed when attacked by Admiral Togo and the Imperial Japanese Navy as they attempted to pass through the Straits of Tsu-Shima during the following May.[4] One of the few Russian vessels that survived the battle was the cruiser *Aurora*, which, many years later, in October 1917, fired the round that marked the opening of the Bolshevik Revolution and is now a museum ship moored in St Petersburg. Back in Britain, trawling interests were later awarded £65,000 in compensation[5] and afterwards a number of new trawlers entering the Hull fleet were given names with a Japanese theme.[6]

Three trawlermen lost their lives as a result of this unprovoked attack and in 1906 a statue was raised on Hull's Hessle Road to mark the event and remember those who were killed. A century after the incident, in 2004, the statue was renovated, and the Russian Naval Attaché attended the rededication ceremony, proffering his country's apologies.[7]

Although the now seemingly distant Russo-Japanese War of 1904–1905 had this dramatic, immediate, and profoundly direct impact on Hull fishermen, the nature and course of the wider conflict, and the lessons learned from it by naval strategists, have been judged as much more significant, with many more long-term implications for the British fish trade and for the fishermen who manned these vessels. This war has subsequently been described as the first modern war. What can more certainly be said is that it was the first twentieth-century conflict in which both sides deployed a great deal of what was then modern naval weaponry – products of the new military-industrial order. Armour-plated warships with large guns engaged each other at what had hitherto been considered great distances. Torpedoes were used in substantial numbers and, above all, both sides made extensive use of mines.

Although mines were then relatively unsophisticated, they were cheap, and used to great effect in this conflict, sinking many warships. Indeed, the threat they posed to warships and merchant vessels alike was evident to all. Although both the Russians and Japanese had deployed them in great numbers, neither side had found a means of effectively neutralising their threat. In the years following this war, international attempts were made to prohibit their use. As these proved unsuccessful, efforts to find to the best ways of countering their threat were redoubled. The need to find a way of dealing with the deadly menace posed by

4 M. Stille, *The Imperial Japanese Navy of the Russo-Japanese War* (Oxford, 2016), 14–15.

5 Rotem Kowner, *Historical Dictionary of the Russo-Japanese War*, 2nd edn (London, 2017), 137.

6 See, for example, Alec Gill, *Lost Trawlers of Hull: Nine Hundred Losses between 1835 and 1987* (Beverley, 1987), 74 and 89–90.

7 The author attended the rededication ceremony.

mines was one of the crucial lessons learned by military strategists from the course of the Russo-Japanese War, and was one key reason why fishermen and fishing vessels came to play a vital role on the maritime front line in the Great War which started almost ten years later.[8]

But, before the Russo-Japanese War, before the Russians opened fire on the Gamecock trawling fleet in October 1904, the idea that the fishing industry had any important part to play on the maritime front line in any future war would have seemed to civilian and military strategists alike almost as improbable as was the Dogger Bank incident itself. Events proved otherwise.

The role that fishermen and the fishing industry subsequently came to play in the Great War has not generally been taken into account in academic naval publications, perhaps at best being considered a marginal part of the war at sea, a niche activity. This book will examine the view that the contribution they made to the British maritime war effort was actually much wider and more fundamentally important than has previously been supposed.

[8] See Chapter 1.

Chapter One

Fish and Naval Forces: The Edwardian Background

The British Fish Trade

In the Edwardian era, by the time of the Dogger Bank incident in October 1904, Britain was not only the world's leading military and mercantile maritime nation, it also possessed the largest and most sophisticated fishing industry the world had ever known. Tens of thousands of fishermen worked from many hundreds of fishing stations – large ports and small villages alike – scattered all around the country's long coastline. The rich waters surrounding the British Isles were home to countless species of fish, many of which were taken by different groups of fishermen in a diversity of locations using a bewilderingly varied range of catching equipment and craft. But although the British fish trade was both complex and considerable it was by this time dominated by two well capitalised and seemingly modern sectors: these were the trawl and herring fisheries, and both employed very large numbers of Edwardian fishermen and fishing vessels.[1]

The Trawling Sector

Trawlers were thus named because they deployed trawls: these were large bag-shaped nets which were towed, or rather dragged, along the seabed. Their catches of white fish – cod, haddock, and the like – were destined primarily for the home market. The British trawling trade had expanded markedly since the mid-nineteenth century when the construction of the national railway network had provided swift and reliable access to inland markets. Railways had transformed the market for perishable foods and in the process helped make

[1] Robb Robinson and D. J. Starkey, 'The Sea Fisheries of the British Isles, 1376–1976: A Preliminary Survey', in Poul Holm, David J. Starkey, and Jón T. Thór (eds), *The North Atlantic Fisheries, 1100–1976: National Perspectives on a Common Resource* (Esbjerg, 1996), 136–137.

fresh white fish an article of cheap mass consumption in many burgeoning inland industrial towns and cities, a much-needed source of protein and a wholesome meal for many working-class families. Somewhere, some time in the following decades, some genius – and no one really knows who – had put fish with chips and created what became a national institution.

Initially, the trawlers deployed by this flourishing trade were wooden-hulled sailing vessels, often owned by the skipper or by very small companies, but from the early 1880s these craft were rapidly displaced by steel-hulled ships fitted with triple expansion steam engines. By the end of Victoria's reign, only Lowestoft in Suffolk and Brixham in Devon retained substantial fleets of sailing trawlers and their replacement elsewhere by the more expensive but far more efficient steam trawlers had been accompanied by a radical reorganisation of this branch of the fish trade. Many of the small owner operators, so common in the sailing trawler era, had gone out of business, displaced by much larger joint stock limited liability steam trawling companies.[2] The divisions between employer and employed, capital and labour that had hitherto been quite imprecise in the fishing industry had become much more clearly defined in the steam trawling sector. Henceforward, the highest position most trawl fishermen could expect to attain was that of skipper; ownership was an increasingly remote possibility for those without access to substantial sources of capital.[3]

The gear deployed by the steam trawlers was also much more formidable than had been the case in the sailing era. These new vessels were fitted with powerful steam deck winches and used wire rather than rope warps to haul the gear. Most sailing trawlers had used beam trawls, so-called because a long and unwieldy wooden beam was used to keep the mouth of the net open, but in the 1890s trawlers began to utilise the much larger otter trawl (Figure 1). The wide mouth of one of these nets was kept open by a pair of otter boards – otherwise known as kites – which stretched it apart when dragged through the water. This system not only made the trawl a far more efficient fish-catching contraption, it also offered many possibilities for adaptation and development which were to prove useful in sectors outside of fishing.[4]

The development of the steam trawler had also led to a concentration of this branch of the trade on the larger ports which possessed the requisite marine engineering, harbour, and bunkering facilities. The two largest British centres were the Humber ports of Hull and Grimsby, while Aberdeen boasted the largest fleet of trawlers in Scotland. Although many of the major steam trawling ports were based on the North Sea coast, Fleetwood and Milford Haven were important

2 Martin Wilcox, 'Concentration or Disintegration? Vessel Ownership, Fish Wholesaling and Processing in the British Trawl Fishery, 1850–1939', in D. J. Starkey and James E. Candow (eds), *The North Atlantic Fisheries: Supply, Marketing and Consumption*, 1560–1990 (Hull, 2006), 56–57.

3 Robinson, *Trawling*, 114.

4 Ibid., 112.

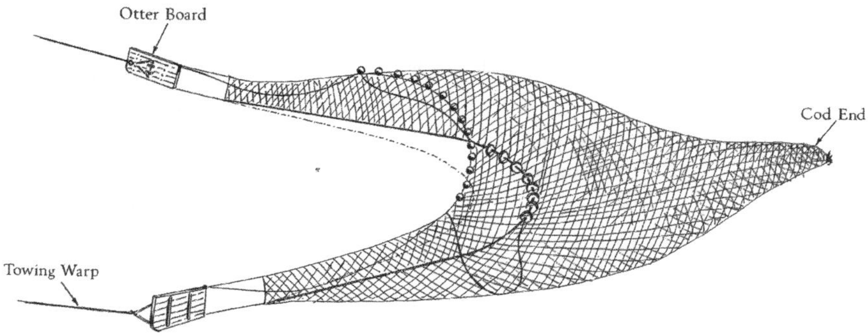

Figure 1: An otter trawl
Source: Peter Mason

west coast centres[5] and there was a small steam trawling centre across the Irish Sea and focused on Dublin.[6]

The introduction of steam was accompanied by a great extension in the geographical reach of the British trawling trade. Not only were the North Sea and other regions adjacent to the coasts of the British Isles now worked more intensively, but distant water grounds, those off the coasts of far-off countries, were being opened up (Figure 2). In the 1890s, steam trawlers started working off the Faroes and Iceland; and around the time of the Dogger Bank incident some vessels were trawling off the Portuguese and North African coasts or else exploring the potential of the Barent's Sea and other Arctic regions. By 1914, the northern distant water grounds were substantial sources of white fish.[7]

Trawl fishing was – and indeed remains to this day – an extreme occupation. In the Edwardian era, trawler crews worked far out to sea on the open, pitching, and exposed decks of small ships, in all seasons and all manner of conditions. Sea-seasoned trawlermen were skilled in handling steam winches and steel warps and in towing and hauling heavy gear as well as in catching, gutting, and packing away fish with the greatest despatch. They toiled, often for weeks on end, far from land, pursuing their fugitive and wild prey across an uncertain and unpredictable maritime wilderness. Theirs was a remote occupation, subject to the capriciousness of both weather and waves and carried out far from the

5 James R. Coull, *World Fishery Resources* (London, 1993), 93–94.

6 Pat Sweeney, *Liffey Ships and Shipbuilding* (Cork, 2010), 58.

7 Robb Robinson, 'The British Distant Water Trawling Industry 1880–1939', in David J. Starkey and Alan Jameson (eds), *Exploiting the Sea: Aspects of Britain's Maritime Economy Since 1870* (Exeter, 1998), 155.

Figure 2: North Sea fishing grounds
Source: Ken Knox

comforts and seeming safety of hearth and home, a way of life quite alien and unknown to those who lived and worked on land, even to many other seafarers.[8] They had to be adept at their job if they were to survive for any length of time. Most experienced trawlermen were, of necessity, very skilled seafarers and this was especially the case with skippers and mates.

A good trawler skipper was at the top of his trade, and the only way to get to his position on the bridge was to start on the lowest rung, as a deckie learner, galley boy, or trimmer.[9] The career path to skipper lay through learning every aspect of fishing other than the technical operation of the steam engines, which was the preserve of the engine room staff. Though skippers and mates were required to demonstrate theoretical competence and acquire appropriate Board of Trade recognised certification – 'tickets' as they were colloquially known on the Humber – their real seamanship skills were learned on the high seas. There were few finer seafarers to be found on any of the world's oceans in the early twentieth century but until this time their unique skillset had not seemed to be of any likely use to the world of the professional, highly disciplined, and thoroughly modern Royal Navy.

The Herring Fisheries

The herring fishery, the other leading branch of the British fish trade, was organised in quite a different way from its trawling counterpart. This fishery had also expanded rapidly over the previous half century and many of the herring caught were salted, packed in barrels, and exported to eastern and central European ports, before being distributed inland by way of the still-expanding continental railway systems; while others were lightly smoked in kilns, processed into kippers or bloaters, and were then despatched by rail to inland markets, and from there to commercial and domestic breakfast tables across the United Kingdom.[10]

Contemporaneously, most herring were caught when they massed together in huge shoals, trapped by their gills in long curtains of nets, known as drift nets, which literally hung in the sea. The vessels deploying these nets were known as drifters and, as was the case with trawling, steam had steadily replaced sail as a means of motive power. However, steam drifters were generally smaller and less powerful than their trawling counterparts. They were fitted with steam capstans rather than the stronger and more versatile steam winches and being cheaper to construct – a fair number had wooden hulls – were still often largely owned by working fishermen rather than large fishing companies (although more members

8 Robb Robinson, *Far Horizons: From Hull to the Ends of the Earth* (Hull, 2010), 54.

9 'Trimmer', otherwise known as a 'stoker'.

10 James Nicolson, *Food from the Sea* (London, 1979), 79–83.

of the crew tended to be hired hands with no direct ownership of either fishing gear or vessel than was previously the case).[11] Although the steam drifter was at the cutting edge of this fishery, there were still many older sailing vessels participating in the Edwardian herring trade.[12]

The herring fishery was a very mobile affair. Various 'races' of herring shoal at different times – known as seasons – in geographically distinct sea areas up and down the British coast, and the fleets of herring drifters followed the seasons voyaging from one to another. The North Sea seasons, for example, started off Shetland in June and finished with the great East Anglian herring fishery in the autumn. As the drifters and fishermen tracked the herring seasons and voyaged down the North Sea coast each year, they were in turn followed by shore-based herring curers and their gangs of processors, mainly women, many recruited from the highlands and islands of Scotland, who carried out the processing of the newly landed herring in curing yards and smokehouses situated quite close to the quaysides where the drifters discharged their cargoes. Speed was of the essence in curing, as herring, being oily fish, are apt to deteriorate rapidly if not processed soon after landing.[13]

Most herring fishermen, like their trawling counterparts, were proficient seafarers, used to contending with all sorts of weather and sea conditions, but their voyages were of much shorter duration and their skills – handling and lying to nets drifting in the sea – were somewhat different. In the off season, in winter, those who turned to white fishing for employment tended not to trawl but to catch white fish by means of longlines set with baited hooks. Herring fishermen were also drawn from a far wider range of fishing communities. Whereas trawling had tended to concentrate both the fleets and its labour force on larger, more urban ports, many drifters and their crews were drawn from smaller fishing communities, in addition to bigger ports such as Aberdeen or Shields, and although herring fishermen came from all British coasts from Cornwall to Shetland the trade was increasingly concentrated in the hands of both Scottish fishermen and curers.[14]

Although the trawl and herring sectors dominated the British fish trade, there were still many other groups of fishermen, typically working inshore waters with small, often undecked sailing vessels and operating from the myriad of small villages and harbours to be found on almost every stretch of the long coastline of the British Isles. Some were more or less full-time catchers of fish, perhaps in popular southern English seaside resorts, supplementing their income with tips

---

11 James R. Coull, *The Sea Fisheries of Scotland: A Historical Geography* (Edinburgh, 1996), 126–127.

12 Chris Reid, 'From Boom to Bust: The Herring Industry in the Twentieth Century', in David J. Starkey, Chris Reid, and Neil Ashcroft (eds), *England's Sea Fisheries: The Commercial Sea Fisheries of England and Wales since 1300* (London, 2000), 190.

13 See Coull, *The Sea Fisheries of Scotland*, chap. 8 and Mike Smylie, *Herring: A History of the Silver Darlings* (Stroud, 2004), chap. 16.

14 Reid, *Boom to Bust*, 188–189.

from taking visitors on sea trips in summer seasons, while others, particularly in the Scottish Highlands and Islands, combined crofting with fishing.[15] The inshore fishing activities, such as the Cornish seasonal pilchard fisheries, had been in relative decline for decades[16] and the activities of the shrinking numbers of such fishermen formed an ever more marginal part of the British coastal economy. But even though the quantities of fish they landed were dwarfed by the catches made in the modern trawl and herring sectors they still constituted a significant element of the fishing industry and an important if dwindling section of the seafaring labour force. Most had an intimate and unmatched understanding of their local waters, in part inherited from the countless generations of catchers who had preceded them, and in part gleaned from years of practical personal seafaring experience.[17]

But, the inshore fisheries notwithstanding, the British fishing industry was a vibrant and substantial part of the nation's maritime economy. Considering others working out of ports on the coasts of Ireland, the Isle of Man, and the Channel Islands, as well as England and Scotland, there were just over 100,000 men and boys active as fishermen in the British Isles in 1913 according to the official reports of the Fishery Boards for that year, and, for most, fishing was their full-time employment. Fishermen formed a huge reservoir of maritime skills and they went to sea in an armada of vessels, there being, excluding the very smallest craft, no fewer than 7,271 fishing vessels working English, Welsh, and Scottish ports alone in 1913. The larger modern steam vessels, of course, formed a much smaller component of this figure but the fleet sizes were still impressive. The combined steam trawling fleet of England, Scotland, and Wales in 1913 was in the region of 1,657 vessels and the equivalent steam drifter fleets boasted a total of 1,555 vessels.[18]

The Fishermen

But to return to the fishermen themselves, it is clear from this brief survey that they were drawn from a wide geographical spread of coastal communities, both rural and urban, large and small, traditional and modern. Yet all had one thing in

---

[15] Coull, *The Sea Fisheries of Scotland*, 187–188. Crofting is a form of subsistence farming practised in Scotland.

[16] Tony Pawlyn, 'Fisheries of the West Country', in Starkey, Reid, and Ashcroft, *England's Sea Fisheries*, 197–198.

[17] Robb Robinson, 'The North Sea Littoral and the British Isles', in David J. Starkey and Ingo Heidbrink (eds), *A History of the North Atlantic Fisheries*, vol. 2, *From the 1850s to the Early Twenty-First Century* (Bremen, 2012), 328.

[18] British Parliamentary Papers, Board of Agriculture and Fisheries, Annual Report on Sea Fisheries for the Year 1913, Cd 7448 (HMSO, London, 1914), 66–67 and 74 and Thirty-Second Annual Report of the Fishery Board for Scotland, Being for the Year 1913, Cd 7399 (HMSO, Edinburgh, 1914), xiv–xv and 169.

common, something that made them quite extraordinary amongst their contemporaries and made them stand out even from their fellow seafarers. It was their occupation. Fishing was, and indeed still is, a unique activity; it cannot be classified as agriculture, orthodox industry, or even transport, though it has affinities to all three sectors. It is the essence of fishing that is quite different: fishermen make their livelihood by pursuing a fugitive prey across a vast maritime wilderness, one devoid of the normal property and ownership structures found on land. It is the process of capture that turns wild fish, a common and unowned resource, into private property. Fishermen in that pre-war era were by then already the last substantial group of people in the British Isles whose essential skills included those of the hunter and who made their living from hunting. During the looming conflict, many would deploy their hunting skills in ways that few fishermen, or their contemporaries, could scarcely have imagined just a few years earlier.

The Edwardian Royal Navy

In October 1904, two days before Russia's Baltic fleet had mistakenly opened fire on unarmed English fishermen in the North Sea, Admiral John (Jackie) Fisher was appointed First Sea Lord.[19] The Dogger Bank incident nearly brought about war with Russia, but Fisher, so new to this crucial post but already a commander with immense experience, advised strongly against taking retaliatory naval action or directly intervening in the war between Russia and Japan. In any case, international tensions eased after Russia agreed to a full investigation of the unfortunate incident and its fleet sailed on.[20]

Fisher was passionate about long-term naval strategy: a mercurial character, he had first seen naval action in the Crimean War and had been at the forefront of many innovations and developments in the decades which followed as he rose through the ranks to the very top of the Admiralty. He took up the role of First Sea Lord with a wide-ranging root and branch plan to completely reform the Royal Navy.[21] Key to these reforms was the development of revolutionary designs of warships – new battle-cruisers and battleships – vessels which Fisher hoped would reinforce Britain's global dominance and see off potential challengers. Within a couple of weeks of being appointed he had set up the Committee of Designs, later known as the Dreadnought Committee. This deliberated while the Russians continued their long and slow voyage to the Far East and by the time

19 Nicholas A. M. Rodger, 'Anglo-German Naval Rivalry 1860–1914', in Michael Epkenhans, Jörg Hillmann, and Frank Nägler (eds), *Jutland: World War I's Greatest Naval Battle* (Lexington, Ky, 2015), 14.

20 Robinson, *Trawling*, 127.

21 Andrew Lambert, 'Tradition, Technology and Transformation', in Robert J. Blyth, Andrew Lambert, and Jan Ruger (eds), *Dreadnought and the Edwardian Age* (Oxford, 2011), 24.

Admiral Rozhestvensky's fleet had been destroyed by Admiral Togo at the Battle of Tsushima on the 26/27 May 1905 the plans for what was to become HMS *Dreadnought* were more or less completed.[22]

Although Fisher had been keen for the Committee of Designs to promote the battle-cruiser it was the battleship for which they initially opted. The new warship was built swiftly: the keel was laid down in Portsmouth Dockyard in October 1905 and the vessel was in service by 2 October 1906. It was a remarkable ship and made an immediate and long-lasting impact. *Dreadnought* contained everything that was cutting edge in battleship technology. An armour-plated, steam turbine, all big gun warship, it could outperform and outgun its predecessors and was said to render all previous warships obsolescent. Lessons learned from the Russo-Japanese War had been taken account of during design and construction, most notably the naval action between their two battle fleets on 10 August 1904, which had demonstrated the accuracy and power of big guns at hitherto unprecedentedly long range.[23] This is said to have ensured that *Dreadnought* was built with 12 inch guns and thus possessed a potential broadside said to be twice that of any other warship.[24]

But Jackie Fisher's new vessel was designed not primarily with the intention of going to war so much as a being a deterrent, a means of underpinning Britain's naval dominance. Although there was a Nelsonian link to every battleship named in Fisher's era as First Sea Lord, and although he often adopted an apparently belligerent attitude, his underlying strategy was based on deterrence, of avoiding the need to go to war to maintain Britain's maritime pre-eminence.[25] By creating such an unprecedented marque of technologically advanced warships, he hoped to extinguish or at least reduce the ambitions of other countries intent on challenging Britain's international naval supremacy. Although dreadnoughts were expensive vessels, they could, so the thinking went, save money if they reduced the ruinous costs of international naval arms races.

Fisher's big ship strategy was not all about dreadnoughts. Anything but. The construction of squadrons of battle-cruisers rather than battleships was pivotal to his original strategic thinking on taking office as First Sea Lord. These heavily armed vessels exchanged some degree of the armour found on dreadnought type battleships for greater speed and manoeuvrability and were originally envisaged not only for protecting Britain's far-flung sea lanes but also for taking on enemy vessels in the North Sea.[26]

---

22 Rodger, 'Anglo-German Naval Rivalry', 14.

23 Roger Parkin, *Dreadnought: The Ship that Changed the World* (London, 2015), 102–103.

24 Richard Hough, *Dreadnought: A History of the Modern Battleship* (Penzance, 2003), 17–18.

25 Lambert, 'Tradition, Technology and Transformation', 24–26.

26 Nicholas A. Lambert, *Sir John Fisher's Naval Revolution* (Columbia, SC, 2002), 9–10.

The overall big ship strategy was successful in the medium term. It is said that the launch of HMS *Dreadnought* brought existing battleship construction across the world to a halt. Some countries either delayed entering the dreadnought race or introduced such vessels in limited numbers. Only the USA and Germany took up Fisher's initial challenge.[27] Germany was certainly wrong-footed. Dreadnoughts were expensive, and they were also too large to use the strategically important Kiel Canal, which provided the German navy with a secure link between the Baltic and North Seas. Nevertheless, Germany took up the challenge, but the costs were prohibitively expensive. It took Germany eight years to widen and deepen the Kiel Canal, which eventually reopened in June 1914, just a couple of months before the start of the Great War.[28] Although many German dreadnought type battleships were built, Britain already had a clear start. By August 1914, there were 22 dreadnoughts and nine battle cruisers in the British Grand Fleet compared with just 15 dreadnoughts and five battle cruisers in the German High Seas fleet.[29]

Furthermore, Fisher's strategy was about more than just capital ships. He invested heavily in the development of the submarine and various other aspects of new technology, not least wireless telegraphy and long-range torpedoes. He realised such innovations had the potential to revolutionise naval warfare and perhaps even the nature of the Royal Navy.[30]

The mid-Edwardian Royal Navy that Fisher took over was clearly a highly professional force, utilising specialist and thoroughly trained personnel in all ranks and positions. Though there was a substantial Royal Naval Reserve (RNR), the high-tech nature of much of Fisher's strategy must have seemed at first sight to have underpinned the need for the Royal Navy to be both a specialist and highly professional service. The impressment of skilled seafarers, including fishermen, so common in early centuries, had effectively ceased over 90 years earlier with the end of the Napoleonic Wars, and after falling into abeyance there was no interest, appetite, or perceived need to try and revive the practice. Yet, when Fisher took up office in 1904, few military strategists can have had much idea about how the need to combat the threat posed by one particular piece of modern maritime military technology, the sea mine, would once again draw so many civilian seafarers and their vessels – most notably from the British fishing industry – into the ranks of the world's most advanced naval fighting machine, albeit in a markedly different way.

---

27 Lambert, 'Tradition, Technology and Transformation', 26.

28 Ibid., 27.

29 Spencer Tucker, *The Great War, 1914–1918* (London, 1998), 49–50.

30 Nicholas A. Lambert, *Sir John Fisher's Naval Revolution*, 10.

Contending with the Sea Mine Threat

Although the design and deployment of armaments on the dreadnoughts was to be influenced by the course of the Russo-Japanese War, naval strategists had also to contend with other outcomes of this conflict, not least the growing effectiveness of mine warfare. The naval mine (or torpedo, as it was sometimes misleadingly called) was not really a new concept: it had been around in one form or another since at least the seventeenth century. In the nineteenth century, mines had been used by the Russians in the Crimean War, by both sides in the American Civil War, and with varying degrees of success in other, later conflicts. Early mines were often unreliable and beset by the problem of effective detonation, but a major step forward came with the development of the Hertz Horn in 1866. This device, invented by German scientist Heinrich Hertz, gave sea mines their characteristic metal horns, inside of which was a glass tube that contained a solution of potassium dichromate. If a ship ran onto a mine then one or more of the horns would break, the solution released by the shattered glass then completed an electrical circuit between two plates, one zinc and one carbon. This created a simple battery with current sufficient to trigger the mine's electric detonator.[31] Mines with Hertz Horns were to send thousands of ships to the bottom of the sea in twentieth-century conflicts.

The Russians initially maintained a keen interest in the development of the sea mine in the ensuing years,[32] but by the early twentieth century several other countries had followed in their wake. The first time that contact mines were used extensively by both sides on the high seas was, of course, in the Russo-Japanese War where these relatively cheap and rather unsophisticated weapons made a crucially destructive impact.[33] Total losses to mines by both sides in this war amounted to three battleships, five cruisers, and three destroyers, along with the lives of thousands of sailors. Vice-Admiral Stepan Makarov, the charismatic commander of the Russian Pacific fleet, also lost his life, along with 26 other officers and 652 sailors when the battleship *Patropavlovsk* succumbed to mines on 13 April 1904, while the warships *Hatsuse* and *Yashima*, one-third of Japan's battleship fleet, were both sunk after striking mines on the same day in May 1904.[34] Whilst both sides clearly used mines with considerable destructive efficiency and demonstrated their effectiveness against all classes of vessel, it was also quickly apparent to observers that neither side had devised a workable means of neutralising their threat.[35]

---

[31] Spencer Tucker (ed.), *A Global Chronology of Conflict: From the Ancient World to the Modern Middle East* (Oxford, 2010), 1328.

[32] Lawrence Sondhaus, *Naval Warfare, 1815–1914* (London, 2001), 168.

[33] Howard S. Levie, *Mine Warfare at Sea* (London, 1992), 17–18.

[34] Rotem Kowner, 'The Impact of the War on Naval Warfare', in Rotem Kowner (ed.), *The Impact of the Russo-Japanese War* (London, 2007), 281.

[35] Norman Friedman, *Fighting the Great War at Sea: Strategy, Tactic and Technology* (Annapolis, Md, 2014), 335.

The need to deal with the threat posed by mines occupied much thought in the ensuing years, but prior to this time very little attention had been devoted by the Admiralty to the problem of finding an effective form of minesweeping. Indeed, sweeping had until then been viewed as a marginal activity, a means of locating a lost torpedo or such like and generally conducted by a junior officer. The lessons learned by the Russo-Japanese War fundamentally shook such perceptions. Afterwards, it was apparent that the effectiveness of a navy's ability to clear as well as lay mines might prove an essential factor in any future war, particularly in the case of Britain with its long coastline, crowded shipping lanes, busy ports, and key strategic naval bases. Yet the sweeping of mines from the sea seemed to be potentially a time-consuming business requiring substantial numbers of skilled personnel and suitable ships.

The need to develop a comprehensive minesweeping strategy was underpinned by international developments. In the wake of the Russo-Japanese War the major naval powers recognised the undiscriminating threat posed by mines to military and civilian vessels in both war and peace, but attempts to bring about an effective ban on their use at The Hague Conference of 1907 proved less than satisfactory, and a further conference held in 1909 emphasised the improbability of securing any effective and realistic international rules regarding their deployment in times of war.[36] As intelligence reports reaching the Admiralty indicated that some countries were ordering the construction of large quantities of sea mines, it became more and more apparent that these weapons would play a major part in any future naval conflict.

Admiral Charles Beresford was amongst a small group of the high-ranking Admiralty officers who wrestled with the problems of neutralising the threat posed by mines in the aftermath of the Russo-Japanese War. Whilst Commander in Chief of the Mediterranean Fleet between 1905 and 1907 he had tried various minesweeping experiments utilising tugs and destroyers, but these had been largely disappointing in outcome and the need to find a solution continued to loom large. On 27 June 1907, shortly after being appointed Commander in Chief Channel Fleet, he visited Grimsby with some of his warships while on a cruise of East Coast ports.[37] The visit was something of a spectacle: decorations and flags welcomed the warships to the trawling town and festivities were organised ashore for both officers and men. The Mayor of Grimsby visited the fleet the following day and afterwards Beresford attended a civic luncheon. During the visit he took the opportunity to inspect a number of trawlers, talk to their skippers and, most notably, hold discussions with Councillor George Alward, one of the most

---

36 National Archives, Kew (hereafter NA), ADM 186/604, 'History of British Minesweeping in the War' (unpublished document dated December 1920), 6–7. See also O. Nippold, *The Development of International Law after the World War*, translated from the German by Amos S. Hershey (Oxford, 1923), 141.

37 *London Daily News*, 28 June 1907.

perceptive and innovative British trawler owners. Alward had played a key role in the development of not only the steam trawler but also the steam winches used on these vessels.[38]

Alward was probably the leading authority on this class of vessel and Beresford was evidently impressed by what he heard and saw. It was clear to him that steam trawler crews were expert at handling wire ropes and steam winches, and towing and hauling in all sorts of weather. There were also some similarities between the methods of trawling and what was thought likely to be good practice in minesweeping. After his visit, he wrote the following:

> Our fishing fleets in war will be rendered inactive, and will in consequence be available for war service. Fishermen, by virtue of their calling, are adept in the handling and towing of wires and trawls, more so than are naval ratings.[39]

Beresford pushed for minesweeping experiments to be carried out using Grimsby steam trawlers and their crews, and his proposal was accepted. There was some slight delay in getting this initiative under way, caused possibly by concerns raised about the status of civilian fishermen engaged in minesweeping during a war, but early in the following year Commander E. L. Booty from Beresford's battleship *King Edward VII* was sent to Grimsby, where he selected two modern Grimsby steam trawlers, the *Andes* and *Algoma*, which were duly despatched south, arriving in Portland on 5 February 1908 with their skippers and usual crews of nine ratings.[40]

Over the next eight days, the trawlers were put through their paces, sweeping up dummy mines under the supervision of Captain F. C. D. Sturdee (later Sir Doveton Sturdee) and the rest of the Channel Fleet Mining Committee. The *Andes* and *Algoma* were not the largest trawlers by then available and were of the kind generally used for working on the North Sea rather than the distant water grounds now increasingly exploited by the British fishing industry off Iceland, the Barents Sea, and other places. They measured no more than 105 feet from stem to stern and had a maximum speed of 8½ knots.[41] What they lacked for in size they more than made up for in terms of robustness and seakeeping qualities, as apparently did their crews.

The first trials used a form of dummy sweep, not dissimilar in many respects to a conventional trawl, but these proved less than satisfactory. Then the otter boards used on a conventional trawl were replaced with wooden prismatic kites

---

38 Robb Robinson, 'Alward, George Lowe (1842–1933), trawler owner', *Oxford Dictionary of National Biography* (Oxford, 2016).

39 NA, ADM 186/604, 'History of British Minesweeping in the War', 3.

40 Archibald Hurd, *History of the Great War: The Merchant Navy*, vol. 1, *1914 to Spring 1915* (New York, 1921), 258.

41 Ibid., 258–259.

and a single wire warp was towed between the two trawlers, its depth regulated by the kites. The experiments showed that by spreading the sweep over two cables in width the trawlers could maintain a seagoing sweep of between 5 and 6 knots, dependent on the weather.[42] According to a 1919 official report on minesweeping, this spread was not improved upon during the course of the later conflict as additional spread increased the weight of the wire and reduced speed. The simplicity and efficiency of the system was soon evident to all involved in the trials and this form of sweep – which became known as the A-sweep – was swiftly adopted and said to have been a means of destroying more than 30,000 mines both during and after the war.[43]

The Committee were impressed not just by the performance of the trawlers and minesweeping gear in the trials but also by their crews who were commended for their enthusiasm, skills, and resourcefulness. There was a further implication evident from these trials: trawl fishermen had a skill set that was in many practical respects more suited to the task of clearing mines from shipping channels than those possessed by conventional Royal Navy officers and ratings.[44]

The Creation of the Royal Naval Reserve (Trawler Section)

The evidence of the Portland trials was enough to convince the Admiralty that trawlers were the way forward and likely to prove indispensable in times of war. Captain Bernard Currey, afterwards Director of Naval Ordnance, suggested that a contract be negotiated and prepared with leading trawler owners that would allow the Admiralty to take up sufficient vessels upon the approach of a war. His idea met with much support, not least from Admiral Sir John Fisher, and work also began on estimating the levels of resources, personnel, and numbers of vessels that would be required to form a suitable and effective minesweeping force.

The next stage in the development of the Admiralty's minesweeping capability was taken forward by Captain Clement Greatorix, who conducted further minesweeping experiments and worked on planning an overall structure.[45] The purchase of six trawlers (two apiece for each of the three Admiralty torpedo schools) was approved so that instruction in minesweeping techniques could proceed as soon as possible. Though there was some irksome delay in obtaining the necessary financial clearance, the first four of these, *Seamew*, *Seaflower*, *Spider*, and *Sparrow*, were acquired in April 1909. They were purchased with very little publicity for what was termed 'subsidiary services'.[46]

---

42 NA, ADM 186/604, 'History of British Minesweeping in the War', 4. One cable = 608 ft. There are 10 cables in a nautical mile.

43 Ibid.

44 Taprell Dorling, *Swept Channels* (London, 1935), 35.

45 NA, ADM 186/604, 'History of British Minesweeping in the War', 4.

46 Hurd, *History of the Great War*, 1:260.

Six gunboat sweepers were also ordered in 1908, and while further experiments took place with destroyers, to develop the fast sweeping practices that would be necessary when working with the main battle fleets, it was also soon evident that a fairly substantial minesweeping Trawler Reserve was to be created. Full details of the new structure emerged in 1910, and envisaged the creation of a minesweeping reserve, consisting of 100 steam trawlers and the enrolment of fishermen of all ranks into a new Royal Navy Trawler Reserve.[47] The Admiralty then commenced negotiations with the Fishery Board and trawler owners to agree terms for the hire of vessels.

The vessels belonging to owners who subsequently agreed to the Admiralty's terms were entered on a specialist list. This meant they had to be released for service as soon as possible after a formal request from the Admiralty. The trawlers were to be hired by the Admiralty at a rate especially agreed for this arrangement. The trawler's hull and outfit were to be valued at £18 for each ton of gross tonnage and at £40 per unit of nominal horse-power and the value thus calculated to be depreciated at the rate of 4 per cent for each full year of the vessel's age. The hire then paid by the Admiralty was to be 12 per cent of this annual valuation. The arrangement also stipulated that the Admiralty had the right to make periodic inspections of all trawlers on the list and, subject to notice, they could remove any they felt fell short of requirements. The owners could buy or sell vessels on the list as they saw fit but had to inform the Admiralty of such transactions, while the vessels on the list could be subject to substitution with the agreement of both sides. Either side could give six months' notice of the intention to end the agreement.[48] Whilst initial planning had envisaged the engagement of 100 vessels, some 146 trawlers were covered by the agreement on the outbreak of war in August 1914.[49]

The initial plans also envisaged the enrolment and training in minesweeping techniques of a reserve of 1,000 fishermen of all ranks who could be called up at short notice to operate these vessels when the need arose. These trawlers and crews were to be formed into units and it was envisaged that each of these units would need to be placed under the charge of an officer, but the problem facing the Admiralty on this account was that, in the event of war, every available officer on the active list was already earmarked for traditional Royal Navy service duties. The solution adopted was to train officers from the emergency and retired lists who were deemed suitable to command units of six minesweepers. In the first instance, 20 such commanders and/or lieutenants accepted the call, and each underwent a 14-day course in HMS *Vernon*: an additional cohort was also

47 NA, ADM 186/604, 'History of British Minesweeping in the War', 4.

48 British Parliamentary Papers, Fisheries England and Wales: Fisheries in the Great War: Being the Report on Sea Fisheries for the Years 1915, 1916, 1917 and 1918 of the Board of Agriculture and Fisheries, Parts I and II, Cmd 585 (HMSO, London, 1920), 20–21.

49 NA, ADM 186/604, 'History of British Minesweeping in the War', 4.

subsequently trained and in a relatively short time the new reserve had a useful nucleus of officers available, all trained in commanding minesweeping units.[50]

With the plans for vessels and officers progressing, it was necessary to recruit the fishermen themselves. It was decided from the start to form a new section of the Royal Naval Reserve for this purpose as many of the men in the existing RNR would be required for service on the bigger ships when the country was on a war footing and, in any case, many, from inshore communities, who did not use trawls, would not have the full set of useful skills possessed by the trawlermen.

Thus, the Royal Naval Reserve (Trawler Section) (RNR(T)) came into existence. When its regulations were formulated in October 1910 it was decided to retain the existing trawler ranks – skipper, second-hand, and the like. The rates of pay were based on normal income expectations in the trawling trade but set some 20 per cent lower. Trawler skippers would have a key role to play in minesweeping operations and those recruited were to be given the rank of warrant officer. To be eligible for enrolment in the new reserve a skipper had to have at least two years' experience of command, possess the appropriate Board of Trade certification, and be no more than 35 years of age. Each had to have undergone eight days of training in one of the Admiralty's small fleet of steam trawlers before receiving the official warrant.[51]

The actual enrolment of personnel, at first planned for 1910, did not commence until early in 1911, when the Admiralty endeavoured to recruit 50 skippers and the same number of mates. Aberdeen was chosen for the first recruitment and training drive, and the six Admiralty trawlers voyaged there, together with two gunboats, the *Jason* and *Circe*, whose commanding officers also had by then some experience of minesweeping. Aberdeen's first minesweeping course for fishermen began on 30 January 1911.[52] Each potential recruit received eight days of instruction which included vessels sweeping independently, receiving the sweep, wheeling and slipping, and sweeping up dummy mines. The first skipper RNR(T) was Peter Yorston, who had been born in North Shields but fished out of Aberdeen. He signed on in February 1911.[53] By the time these Aberdeen sessions were completed in mid-April 1911, some 28 skippers, 27 mates, 20 deckhands, 21 engineers, and 20 trimmers were trained and recruited.[54]

Although this first phase of enrolment and training of Aberdeen trawlermen had proved fairly successful, the drive met with a much less enthusiastic response at the Humber ports and elsewhere. In part this seems to have been due to misgivings about what enrolment actually meant. It was suggested in parliament that some fishermen were worried that by enrolling they would have made

50 Ibid., 4–6.

51 Ibid., 6–8.

52 *Aberdeen Press and Journal*, 31 March 1911.

53 Peter Yorston Service Records, NA, BT 377/7/124241 and NA, BT 377/7/126646.

54 Hurd, *History of the Great War*, 1:262.

themselves liable to serve on other naval vessels and it was also suggested that the age limit on skippers was a key issue: the initial enrolment criteria required trawler skippers to be 35 years old or younger. Many of the really experienced Humber skippers were older than this but the Admiralty had hoped to recruit from the sizeable cohort who were in the right age range. Later, the age limit for enrolment was raised, but in the case of the Humber ports it took longer for the recruitment message to get across. Initially, at Grimsby, no more than a dozen trawlermen volunteered and none was a skipper. Part of the problem was probably that many Humber trawlers were usually at sea for longer periods than their Aberdeen counterparts, so the pattern of their trawler trips did not fit easily with the eight-day training session. However, it was clearly important to win over skippers and key players. Captain Clement Greatorex therefore spent some time at Grimsby talking directly to officials from unions and associations representing skippers, deckhands, and engineers and gradually enrolment prospects improved.[55]

Recruitment continued through the summer and into the autumn of 1911. Fleetwood and Milford Haven, as well as Hull and Grimsby, were visited as part of the process and altogether a further 526 trawlermen of all ranks, including 52 skippers, were trained and enrolled during this period.[56] Despite the slow start, the nucleus of what was to become a great auxiliary navy had come into being.

By November 1911, the future of the minesweeping service was sufficiently clear that a Minesweeping Section of the Admiralty was created and Captain Richard M. Harbord was appointed as Inspecting Captain of Minesweeping Vessels. His many and varied duties included inspecting all vessels taken into the reserve, visiting fishing ports to liaise with trawler owners, overseeing the various training programmes and mobilisation in time of war, as well as keeping acquainted with defence plans, particularly relating to mine clearance in strategically important channels, and perfecting the efficiency of minesweeping arrangements.[57]

As this now substantial reserve force of vessels and men began to take shape, the Admiralty began to formulate plans for its rapid mobilisation. From November 1911, mobilisation officers were appointed, one for each of the main trawling ports, and their instructions and responsibilities were clearly outlined to them. When war loomed, each mobilisation officer would be ordered by telegram to proceed with all haste to their allotted port. On arrival, a second telegram would contain orders regarding the number of trawlers to be taken up and the mobilisation officer would then warn the port's Registrar of Shipping and Seamen to prepare crews to take up the trawlers. In turn, the Registrar was expected to provide the mobilisation officer with an up-to-date list of both trawlers in port

55 Hansard, HL vol. 8, cols. 995–1000 (30 May 1911).

56 Hurd, *History of the Great War*, 1:263.

57 NA, ADM 186/604, 'History of British Minesweeping in the War', 6–8.

and those expected to arrive soon. To ensure such procedures worked effectively, each mobilisation officer was required to undergo a three-day training course in their respective port, become acquainted with the locality, and forge links with important local officials, trawler owners, and harbour authorities. They were also expected to spend one day a year at the Admiralty and confer with the Inspecting Captain of Minesweeping.[58]

Arrangements were then to be made to prepare these vessels to return to sea, but on Admiralty Service. They were to be coaled, watered, and provisioned for seven days while the owners were expected to remove the catch, the ice, and all the fishing gear, except the warps. As the crews came on board and preparations were completed, the mobilisation officer was to provide each skipper with the requisite charts and sailing orders and the vessel would then be expected to voyage at full speed for its allotted destination.

On arrival, the skipper had to draw the newly requisitioned vessel's special minesweeping stores, which included kite, white ensign, cone, and signal book. He was also informed of the sweeping group and parent ship his trawler was allocated to, and, of course, details of the officer in charge. A number was also given to the vessel and this was to be painted on each bow in figures two feet long with the civilian fishing numbers painted out. The vessel was subsequently referred to by its Admiralty number while in service. Trawler skippers and crews were clearly not familiar with naval practice and therefore arrangements were made for each vessel to be assigned a naval petty officer, to be ranked third in command; his duties were to provide the skipper with advice on signalling, account keeping, and all manner of naval matters. By this time, the minesweeping trawler, by then painted from stem to stern in naval grey and flying the white ensign, would be considered to have completed its transition from civilian fishing vessel to a man of war.[59]

In March 1912, with these plans formulated, the retired naval officers who had already been selected were allocated to take command of the minesweeping groups of six trawlers. Their priority, on receipt of their mobilisation telegram, was to travel to their respective port and proceed to sea with their minesweeping group. By September of that year, there were already 64 trawlers on the Admiralty list and each was allocated to one or other of the minesweeping groups, which by then potentially covered – albeit thinly – much of the North Sea and Channel coasts of the British Isles and up the west coast as far as Milford and Queenstown in Ireland.[60]

By the end of July 1914, about 145 trawlers and crew members were enrolled in this new auxiliary force. These fishermen were now ready to be called up for war.

58 Hurd, *History of the Great War*, 1:265.

59 Ibid., 264.

60 Ibid., 266.

Chapter Two

1914:
The Early Months of the War

The First German Minefield off the British Coast

On 29 July 1914, with Europe in the inexorable grip of a deepening and rapidly deteriorating crisis, Winston Churchill ordered the vessels of what was soon to be known as the Grand Fleet northwards from the south coast of England to take up their war station at the remote anchorage of Scapa Flow.

They left in darkness, 18 miles of heavy warships, dreadnoughts, battlecruisers, and their escorts, an immensity of armour plate; sailing at high speed through the Straits of Dover and into the North Sea, northwards, towards their new base in the Orkney Islands, just beyond the top of the Scottish mainland, passing through waters newly swept by the only flotilla of minesweeping gunboats the Royal Navy then possessed. By the time this huge fleet reached Scapa on 2 August, Germany was already at war with Russia and would declare war with France the next day. Within two days of the fleet's arrival in Orkney, Germany would invade Belgium, and Britain would have been drawn into the conflict.

In theory, the giant anchorage at Scapa was well positioned to enable the Royal Navy's fleet of dreadnoughts to command the North Sea and anticipate threats from its somewhat smaller German counterpart, the High Seas Fleet. But although plans to use the anchorage had been in place for some years they had not been backed by the requisite protective expenditure, and this expensive array of dreadnoughts, soon under the command of Admiral Jellicoe, found themselves in a base blessed with very little, indeed almost nothing, in the way of modern defence works.

At 7.30 p.m. on 4 August 1914, just a few hours before Britain formally declared war on Germany, Commander Biermann on the *Konigin Luise*, a small Hamburg America Line excursion steamer, received orders by wireless to proceed to sea from the River Ems. Although the height of the summer, this was to be no excursion trip: no tourists thronged the little steamer's decks. Far from it: the vessel was packed with mines. Captain Biermann had orders to head for the Thames at top speed and dispose of this deadly consignment as close as possible to the shipping lanes that stretched along the neighbouring east coast of England.[1]

[1] E. Keble Chatterton, *The Auxiliary Patrol* (London, 1923), 14.

The vessel's yellow funnels and white hull had been hurriedly painted black to make it look something like one of the Great Eastern Railway civilian steamers running between Harwich and the Hook of Holland. The *Koenigen Luise* shadowed the Dutch coast until reaching the Maas Lightship then struck out on an eastwards course, sailing at full speed for East Anglia. Some time afterwards, the light cruiser HMS *Amphion* and a flotilla of destroyers, alerted by a trawler report that a suspicious looking vessel had been seen casting objects over the side about 30 miles eastwards of Orfordness, sighted the *Koenigen Luise*, by then preparing to return home. A chase ensued, Captain Biermann tried to escape by running first southwards then south-east, but the faster ships of the naval flotilla closed in and around 11.15 a.m. opened fire. Biermann soon scuttled his badly damaged ship, sending a wireless message back to Germany saying that the minelaying had been completed. Forty-six of the minelayer's crew were lost in the action. HMS *Amphion* picked up the German survivors but early the following morning struck the German mines, becoming the first Royal Navy ship to sink in the war, with the loss of 149 members of the crew[2] as well as many of the German sailors who had been taken on board.[3]

The *Koenigen Luise* had laid the first minefield of the Great War and the implications were clear: the British could expect the Germans to pursue an aggressive as well as a defensive mining policy. Coastal shipping was immediately disrupted and although this minelayer had been eliminated the Royal Navy had also lost a light cruiser in the action while the passage of shipping was disrupted up and down the coast. This opening action of the North Sea war underlined the urgent need for the Royal Navy to mobilise and indeed expand its minesweeping capability.

Mobilisation of Trawlers and the Creation of the East Coast War Channels

On the same day that the Grand Fleet had been ordered north, the Inspecting Captain of Minesweepers also received instructions to mobilise the Royal Naval Trawler Reserve while the Admiralty's own small flotilla of minesweeping trawlers, most recently employed in target towing, were ordered to fit out for sweeping at the Nore. On 6 August, suitably equipped, they put back to sea and began to tackle the *Koeningen Luise*'s mines, sweeping from Orfordness to Southwold.[4] Meanwhile, up and down the coast, at all the assigned trawling ports, the mobilisation officers started to implement the Admiralty's minesweeping strategy. Within a short space of time, 92 trawlers out of the 142 vessels on the

2 Commonwealth War Graves Commission, HMS *Amphion*, August 1914.

3 Chatterton, *Auxiliary Patrol*, 14.

4 NA, ADM 186/604, 'History of British Minesweeping in the War', 9.

reserve list had been commissioned into service, and another 100 additional vessels, described as Special Reserve, were ordered to be requisitioned.[5] But from the onset it was evident that the Admiralty would have to take up many more vessels than were covered by the pre-war reserve arrangements if Britain was to have any chance of containing an aggressive German mining strategy.

Within three days of the outbreak of war, a plan to create what was hoped would be a safe channel for merchant vessels from the Outer Dowsing to the South Goodwins took shape, the idea being to deploy a flotilla of 80 trawlers with the intention of keeping a passage through these waters constantly swept and thus always, in theory at least, clear of mines. This alone would require the services of many vessels and Captain Ellison, then commanding officer of the *Halcyon* at Lowestoft, was instructed to create this new minesweeping fleet. He worked quickly and by 11 August the first trawlers deployed had already swept a channel four cables wide from the North Foreland to Southwold. The seemingly urgent need to create and maintain such a substantial swept channel made it imperative to take up many more steam trawlers as rapidly as possible, and indeed move to taking up other kinds of fishing vessels. During the early days of the conflict not only were many more steam trawlers rapidly hired but the less powerful steam herring drifters and their crews began to be drafted as well. By mid-August, this section of what soon became known as the East Coast War Channel was being swept every day by minesweeping trawlers, and continuously patrolled at night by East Anglian steam drifters, who at first had little more than the White Ensign in the way of a deterrent for would-be minelaying vessels. By 29 August, the full inshore channel was not only reported to be clear but had been marked by buoys, and details of this new passage were being forwarded to all ships.[6]

During this first frenetic month, when almost every piece of flotsam was being reported as a mine,[7] steam fishing craft were being hired almost as fast as they could be fitted out for war service. Initially, Lowestoft was the main port dealing with the vessels, converting them to men of war at the rate of six trawlers per day. Local firms of shipwrights and engineers were reported to be at their wits' end by the demand for their services, but soon other ports were drafted in. Besides the East Anglian herring vessels, unarmed steam drifters from the ports of north-east Scotland were also soon being despatched to sea in search of enemy minelaying vessels, and yet more trawlers were being requested for minesweeping duties along the Scottish coasts and in the approaches to naval bases. By 1 September, the numbers of trawlers and drifters taken into Admiralty service already exceeded 250 vessels,[8] almost double the strength of the pre-war

[5] Ibid.

[6] Ibid., 9–10.

[7] Ibid., 9.

[8] Julian Corbett, *History of the Great War: Based on Official Documents, Naval Operations*, vol. 1 (London, 1920), 161.

Photograph 1: Trawler crew on Admiralty service
Source: Edward Keble Chatterton, *The Auxiliary Patrol* (London, 1923).

Trawler Reserve: a new, and hitherto unanticipated naval force of some considerable size was rapidly emerging (Photograph 1).

Apart from patriotic fervour, there was another strong impetus for the owners and crews of fishing vessels to support the Admiralty's rapidly evolving plans. At the outbreak of war, the Government had instructed North Sea fishing to cease and all vessels to return to port. The various North Sea fish trades were immediately and completely disrupted, fish supplies dried up, and those involved were left without work. Although these draconian orders were soon modified, and fishing permitted to resume, albeit governed by a complex web of Admiralty regulations, it was clear that the herring trade, heavily reliant on exporting its produce to markets in central and eastern Europe, was facing huge problems. The spectre of long-term unemployment loomed. Given such uncertainty, many fishermen, particularly those engaged in the herring trade, who had not previously joined the Trawler Reserve, now had a significant economic impetus to take up the offer of using their seafaring skills on war service with the Trawler Reserve.

The Next Phase of North Sea German Minelaying

Further aggressive minelaying by the enemy only three weeks after the outbreak of hostilities seemingly emphasised the immensity of the task facing the officers and men by then being rapidly recruited into the Trawler Reserve. On 24 August, the *Albatross*, a small German cruiser, with the capability of carrying 400 mines, left Heligoland Bight, crossed the North Sea, and began laying mines in a zigzag fashion in an area the commander thought was about six miles seawards of the Tyne. His reckoning was wrong, and the mines were pitched overboard from his ship around 30 miles off shore, but, nevertheless, they still posed a new and deadly hazard to navigation.[9]

In the early hours of the next day, another German minelayer, the *Nautilus*, left the Ems and headed for the Dogger Bank before altering course as it closed on English North Sea waters. As midnight approached, this vessel laid two, quite lengthy, legs of mines over a substantial stretch of water some 50 miles to the eastwards of the Humber and Flamborough Head.[10]

North Sea fishing had by this time resumed and both the *Albatross* and *Nautilus* had been accompanied for part of their voyages by other German naval vessels, which captured and destroyed several steam trawlers and drifters that they encountered, taking their civilian crews of fishermen as prisoners. It seems likely that the Germans had also reconnoitred further potential mining areas in the North Sea, as a couple of days earlier several other trawler crews had also been captured and their vessels sunk. Altogether, 14 Grimsby and ten Boston trawlers were despatched to the bottom in this fashion between 22 and 26 August.[11]

These two new minefields soon made their deadly mark. On 26 August, several drifters fishing off the Tyne found mines which exploded in their nets. A couple of Danish fishing vessels also blew up, along with a Norwegian steamer, and were followed to the bottom by the steam drifter *Barley Rig* and five of her crew. As the carnage mounted, four minesweeping trawlers were sent out from the Tyne around 5.30 a.m. the following day and started trying to clear this yet indeterminate menace of a minefield. It was almost certainly the first time that their crews had attempted minesweeping in hostile conditions, and their inexperience showed. Before the end of the first day, two of the minesweeping trawlers, *Thomas W. Irvin* and the *Crathie* had been blown up by the mines they were trying to remove. The first to go was the *Thomas W. Irvin*: the trawler was involved in the sweeping up of seven mines that day but at 4.25 p.m., while preparing to connect with another vessel for a further sweep, the trawler collided with a mine and blew up. Just over half an hour later, at 5.06 p.m., the *Crathie*, which had already snagged a couple of mines on its sweep wires during the afternoon without incident, sank quickly, when another device detonated beneath its hull. These ships were the first of many minesweeping trawlers to be lost during the conflict and their abrupt end illustrated something of the dangers now facing the Trawler Reserve crews. The survivors from the *Crathie*, who were picked up by a torpedo boat and landed at North Shields, reported that their vessel had sunk

---

9 Chatterton, *Auxiliary Patrol*, 28.

10 NA, ADM 186/604, 'History of British Minesweeping in the War', 10.

11 HMSO, *British Vessels Lost at Sea, 1914–1918* (London, 1919), 99–100.

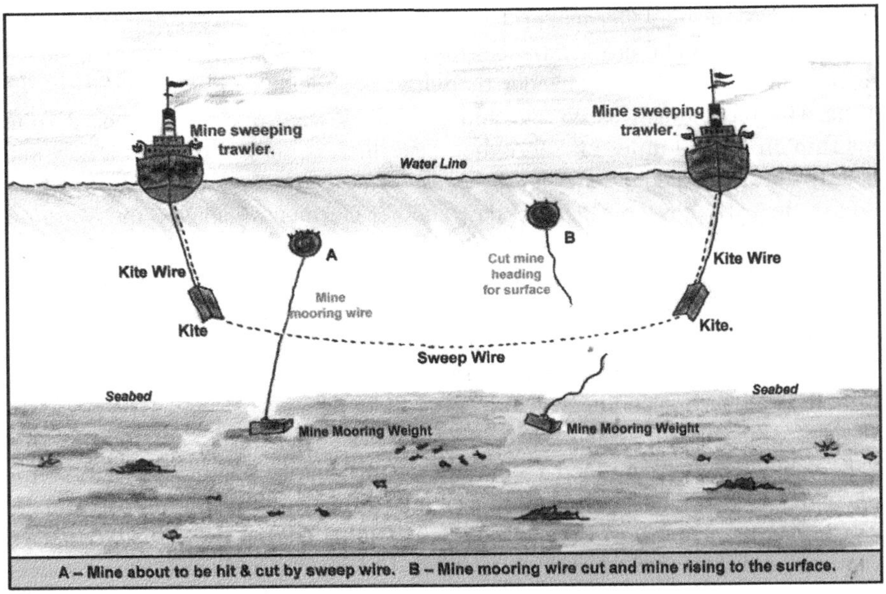

Figure 3: 1914–1918 minesweeping operation
Source: Ken Knox

like a stone – without even time to launch the boat – as the mine exploded, those on deck being blown into the sea.[12] Two crew members were lost with the trawler, their bodies never recovered, and most of the survivors sustained injuries in the bloodbath, with the skipper, Herbert Cook, suffering a compound leg fracture and general bruising. Three members of the crew of the *Thomas W. Irvin* were also lost with their trawler. Less than a month earlier, the crews of both vessels had still been engaged in their normal fishing duties, working mostly out of either Aberdeen or the Tyne.[13]

Six vessels had been lost within 24 hours; and at that point it was decided to abandon attempts to remove the whole minefield and focus efforts on sweeping a safe channel for shipping much closer to the coast. It later turned out that almost 200 mines had been laid in this area by the *Albatross*, and henceforward they were placed out of bounds and left largely alone. This minefield was marked but not fully swept until after the end of the war, more than four years later (Figure 3).[14]

---

12 *Aberdeen Press and Journal*, 29 August 1914.

13 See, for example, Commonwealth War Grave Commission Records, George Crossley Norris, 207DA.

14 NA, ADM 186/604, 'History of British Minesweeping in the War', 9–10.

The field laid off the Humber – which contained a similar number of mines – was also discovered on the same night as its Tyne counterpart. Fishermen again made the discovery: this time the Yarmouth drifter, *City of Belfast*, found two mines amongst its nets while working north-east of the Outer Dowsing Lightship and yet another one exploded in her nets the following day. Two vessels sent to try and clear or at least ascertain the extent of the minefield were lost before the end of the month and soon afterwards a couple more craft were destroyed while fishing in the vicinity.

Alternative and swifter means and forms of sweeping were soon sought. One suggestion put forward was to try deploying conventional drift nets usually used for catching herring to snare mines. Several herring drifters were hastily commissioned, along with their fishing equipment, and initially attempted to ensnare mines close to the Outer Dowsing, where a number of vessels had already sunk. The drifters had some success – a few mines were destroyed – but the system proved complicated and was deemed too dangerous to be of much practical value. Two steam drifters, the *Eyrie* and *Lindsell*, as well as the gunboat *Speedy*, formerly a fishery protection vessel, were sunk in the trials when working about 40 miles off the Humber.[15] One crew member from the *Speedy* died and 11 fishermen, including both skippers, were lost when the drifters blew up. The body of Charles Woodgate, skipper of the *Lindsell* was identified by his wife after washing ashore some time later at Happisburgh on the north-east coast of Norfolk.[16] He was subsequently buried in St Peter's churchyard in the village of Carlton Colville. The others have no grave but the sea.

Admiral of Minesweepers

On 1 September, Rear Admiral Edward F. B. Charlton took up the position of Admiral of Minesweepers, his duties being to co-ordinate minesweeping activities along the east coast of England.[17] By the time he assumed command, an average of one ship had been destroyed for every two mines sunk and the carnage they could create amongst those vessels tasked with their removal was already apparent: a member of a crew of minesweepers was said to have been lost for every mine destroyed. The deadliness of this seaborne device was all too evident, as was the inexperience of officers and crews.[18]

Admiral Charlton voyaged to Grimsby in the yacht *Zarifah* – the deckhands on this newly requisitioned vessel all being Cambridge undergraduates and watermen rather than experienced mariners – then took a flight on one of the

15 Ibid., 10 and *Western Daily Press*, 7 September 1914.

16 *Thetford & Watton Times and People's Weekly Journal*, 31 October 1914.

17 NA, ADM 186/604, 'History of British Minesweeping in the War', 10.

18 Ibid., 11.

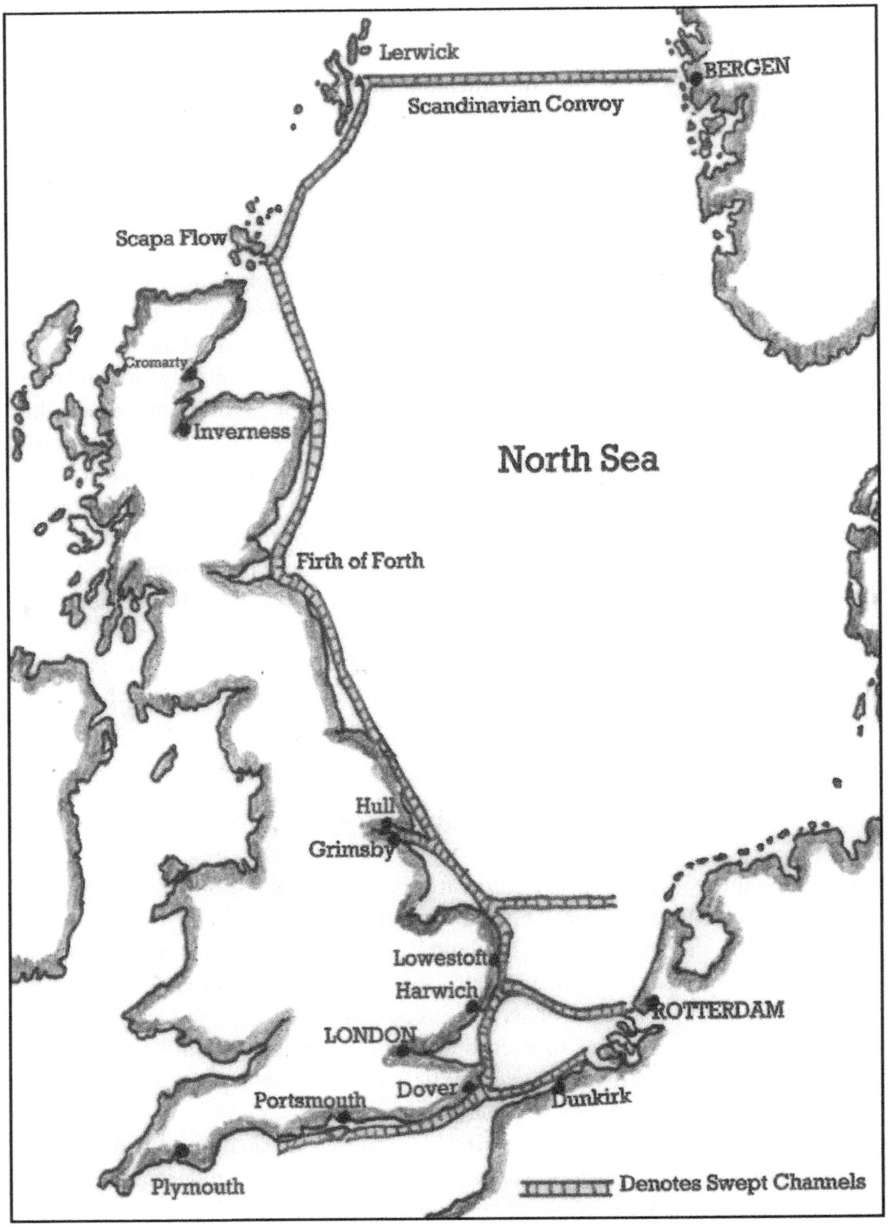

Figure 4: East coast swept channels
Source: Ken Knox

seaplanes based at Killingholme on the Humber to ascertain what he could see of this new Humber minefield. Charlton saw at first hand that the sandy waters of the North Sea made it difficult to spot mines from the air, never mind from the surface of the sea.[19] Winter was approaching, and, as was the case with the Tyne, it was decided to leave the minefield in place and concentrate on maintaining the swept channel closer to the coast.[20]

By mid-September, the east coast swept channel was lined with buoys at an average of one every two miles along its entire route. This constantly cleared sea lane now stretched from the Downs, past Kentish Knock, Longsands to the Shipwash, then onwards by Corton Cross Sand, Newarp and Haisborough, before reaching Cromer Knoll. From there the channel continued even further up the coast, passing the Outer Dowsing and reaching northwards as far as Flamborough Head. In October, some alterations were made to the section of channel originally buoyed between Haisborough and the Spurn Light. The new channel line required vessels to hug the coast more closely, which made it easier both to control shipping movements and to avoid the hazards of the Humber minefield. No other significant further changes were made during the remainder of 1914.[21] This buoyed passage, which become known as the East Coast War Channel, had to be constantly swept and re-swept and also perpetually patrolled in order to deter or disrupt enemy minelaying activity (Figure 4).[22] The fishermen and fishing vessels recently taken up by the Trawler Reserve were tasked with much of this arduous and unending work.

The Early Impact of U-Boats

The need to counter and neutralise aggressive enemy mining was just one of the lessons learned from early events in the war at sea; another was the need to respond to the threats now posed by submarines. Prior to the Great War, the submarine had been perceived by many key military players on both sides as a largely defensive weapon with only a limited range, but from quite early in the conflict it became evident that its potential impact had been considerably underestimated.

On 6 August, ten German U-boats left base on their first war cruise in the open seas looking for the British battlefleet. Whilst one turned for home two days later with engine trouble, the others continued with their quest, and off Fair Isle the U-15 made its initial attack, an unsuccessful attempt to torpedo the battleship

19 G. H. P. Mulhauser, *Small Craft* (London, 1920), 7–8.

20 NA, ADM 186/604, 'History of British Minesweeping in the War', 12.

21 Ibid.

22 For the a substantive account of the east coast war channels and their significance, see Antony Firth, *East Coast War Channels in the First and Second World War*, A Report for English Heritage (Tisbury, 2014).

Monarch. The next day this U-boat paid a heavy price for its audacity when it was spotted on the surface, evidently under repair, by HMS *Birmingham*, which subsequently rammed and sank the enemy vessel. Although another craft, U-13, also failed to return to Germany from this cruise, the attack had demonstrated that U-boats had a far greater operational range than either side had previously anticipated.[23] During the next few weeks, thanks to the early voyages of German Commander Otto Hersing, something more of the full potential of U-boats became evident to both sides. On 14 August, he embarked on a sweep of the North Sea to the Scottish coast, covering 1,600 miles without any significant problems. On another voyage he sailed into the Firth of Forth as far as the Forth Railway Bridge, then three days later made history by sinking the cruiser HMS *Pathfinder* off St Abbs Head: 259 crew members were lost when the stricken warship went down.[24]

At first, the British thought that the *Pathfinder* was the victim of mines,[25] but it became clear that a U-boat was the culprit when no such devices were found in the area. Within weeks the threat posed by U-boats to surface warships was starkly reinforced: on 21 September, Commander Otto Weddigen of the U-9 sank three ageing British battlecruisers, *Aboukir*, *Cressey*, and *Hogue* in shoal waters near Scheveningen. Weddigen's orders had been to harry troopships crossing to Belgium, but he had been driven off course by bad weather and compass problems. On surfacing shortly before dawn, he saw the three British cruisers and attacked. His first torpedo struck the *Aboukir* about 6.30 a.m. and the large warship sank in no more than 25 minutes. The *Hogue*, believing a mine to be the culprit, closed in to pick up survivors, but was in turn hit by two torpedoes and went under within ten minutes. The *Cressey* had also hove-to for survivors and fell victim to two further torpedoes from the U-9, sinking in only 15 minutes. Although Dutch steamers and a couple of Lowestoft trawlers rescued many survivors from the sea, 62 officers and 1,073 men were lost in these ships.[26] The U-9 then returned to its base without continuing its original mission, having used up all its torpedoes.[27]

The Deployment of Fishing Vessels and Fishermen in Anti-Submarine Activities

Even though the *Aboukir*, *Cressey*, and *Hogue* were old ships, the lesson was evident. Three large warships had been sent to the bottom in an hour by one small

---

23 R. H. Gibson and M. Prendergast, *The German Submarine War* (London, 1931), 3–5.

24 Ibid., 5–6.

25 Corbett, *Naval Operations*, 1:163.

26 Gibson and Prendergast, *German Submarine War*, 5–6.

27 Chatterton, *Auxiliary Patrol*, 20.

U-boat. The ramifications of the September warship losses were tremendous. The U-boat had more than proved its worth and the Admiralty now had to find a way of trying to counter the hitherto under-anticipated level of threats posed by both mines and submarines, threats not only to its expensive fleet of sophisticated surface warships but also, as became increasingly evident as the war progressed, to the country's very lifeblood: its maritime trade and mercantile shipping. In both cases the Admiralty's response to these crises included the recruitment of ever more fishermen, fishing vessels, and other small vessels for the maritime front line.

Whilst the Admiralty had expected early torpedo attacks on its large vessels in the event of war, it had anticipated that most would be made by surface vessels, principally fast destroyers. The early weeks of the war showed that offensive minelaying by surface vessels and torpedo attacks by submarines were likely to be commonplace. This was clear from early sightings,[28] and it was apparent that the U-boats had been investigating the fleet's base at Scapa – an area well known to the German navy which had made several peacetime visits to Orkney and Shetland over the preceding decades. Given the limited defences in place at Scapa, Admiral Jellicoe moved his heavy warships further west for some time, establishing a base at Loch Ewe on the Scottish mainland in Wester Ross. And scarcely a fortnight after the war had begun the Admiralty created the Northern Trawler Flotilla. This was based at Scapa and consisted of 16 trawlers, manned by members of the Trawler Reserve, each fitted with an explosive modified sweep and a couple of 3-pounder guns. The trawlers were fitted out at Lowestoft and the first vessels arrived at Scapa before the end of August.[29] As Corbett put it: 'In all tradition it had been a constant duty of the Grand Fleet to protect our fishing fleets; now it was the fishing fleets who must protect our Grand Fleet'.[30]

These vessels were not intended for minesweeping, as had been the anticipated pre-war plan for requisitioned steam trawlers, they were to be used for hunting and harrying submarines off the eastern Orkneys. Fishing vessels – steam trawlers and drifters – were thus already being taken up by the Admiralty to chase submarines and minelaying vessels. By the end of the first month of hostilities it was clear that this war at sea was going to be a conflict in which small vessels had a major role to play. In early September, the Admiralty further extended its activities aimed at combating enemy minelaying and submarine operations. A decision was made to take up all available trawlers, motor boats, and steam yachts and then deploy them in operational units. Each unit was to comprise a yacht, four trawlers, and four motor boats. Many of the yachts taken up had hardly finished their summer cruising season, but all such serviceable

[28] Hurd, *History of the Great War*, 1:333.

[29] Chatterton, *Auxiliary Patrol*, 21.

[30] Julian Corbett, *History of the Great War: Based on Official Documents, Naval Operations*, vol. 2 (London, 1921), 2.

craft were sent to either Portsmouth or Devonport. Wireless equipment was installed in these yachts, which were repainted in regulation grey and armed with 3- or 6-pounder guns mounted fore and aft. In the meantime, 3- or 6-pounder guns were also installed on the trawlers taken up for patrol duties, and all such vessels were subsequently fitted with modified explosive sweeps. The newly formed units were despatched for patrol duties off key sections of the British coasts.[31]

The first units were sent to Scapa, Loch Ewe, Rosyth, Humber, and Cromarty, but as more vessels became available during the rest of 1914 further units were formed. These armed units were expected to patrol given stretches of the coast and were charged with attacking and harrying U-boats and arresting any vessel suspected of laying mines, in addition to other ship inspection and patrol duties. Before September was out, fifteen yachts had been commissioned and a similar number were being fitted out. By this date more than 300 trawlers and 100 motor vessels were already in Admiralty service, either minesweeping or on the new patrol duties, and the numbers were growing by the week. The pace at which commissioning took place stretched every resource: particularly the procedures for surveying and inspecting vessels. It proved particularly difficult if not an impossibility in many cases to ensure all vessels taken up, straight from fishing, were in what was the Royal Naval perception of top condition when entering Admiralty service. Fishing was a hard business; most working trawlers of any age were subject to considerable wear and tear and, while many engines were found to be in satisfactory condition, it was not unusual or perhaps surprising to find reports by mechanics that the all-important winches were often found to be considerably worn.[32] Even before the middle of August, and the appointment of engineering officers to assist with the commissioning process, it had been evident that naval peacetime standards for taking up vessels would have to be compromised. As an Admiralty telegram of the time spelt out, the fact that trawlers were 'being employed in fishing became sufficient warrant for them to be in good enough condition for the work of sweeping', and a similar approach seems to have been adopted for those being utilised for patrol work.[33] Even so, a few of the vessels taken up appear to have been in just too poor condition for any Admiralty duties (not least the trawler *Cameo*, which was considered so unfit for service after arriving at Dover that it was returned to its home port of Milford Haven and the crew were transferred to another requisitioned vessel). In fact, the *Cameo* proved to be in such a bad state that it could only make the voyage back to its Milford owners after what seems to have been significant repair work.[34]

31 Hurd, *History of the Great War*, 1:324–325 and 333.

32 NA, ADM 137/976, Auxiliary Patrol Trawlers, 6 August 1914.

33 Ibid., 11 September 1914.

34 Ibid., 6 September 1914.

Intensification of U-boat Activity

During October, it became all too evident to the British that the Germans were widening and intensifying their deployment of U-boats. Many different types of Royal Navy vessels reported being attacked, not least the cruiser *Hawke*, which was torpedoed and sank off the north-east coast of Scotland on the 15th, while the aircraft ferry vessel HMS *Hermes* sank on the 31st, and there were numerous reports of submarine sightings made on different sections of the coast. On 26 October, the SS *Admiral Ganteaume*, a ship packed with Belgian refugees, was torpedoed by U-24 off Cape Gris Nez, and although the damaged vessel was towed into Boulogne some 40 people on board lost their lives in the chaos which followed the attack. Another ominous development was the stopping and sinking of SS *Glitra* off the Norwegian coast on 20 October by U-17. The *Glitra* was the first of hundreds of British merchant vessels to succumb to U-boat depredations during the Great War.[35]

The capture of the Belgian ports of Ostend and Zeebrugge in October 1914 was to prove of great value in the rapidly evolving U-boat war. The Germans found the harbour facilities intact when they occupied both towns, and the following month the first U-boats arrived at Zeebrugge. The Germans were to make effective use of these strategically placed ports as bases for U-boats and surface craft for much of the remaining four years of the war.

All efforts to counter the U-boat threat by laying defensive minefields around strategic ports and naval bases were hampered by the poor quality of British mines. At the start of the war British manufactured mines were considerably inferior to their German counterparts in every respect. Indeed, only the Germans and Russians then possessed any substantial quantities of mines that would be considered state of the art technology. The limited number of mines manufactured in Britain pre-war used a combination of firing lever and cocked spring-driven striker for the detonation mechanism, which proved less reliable than the mechanism on enemy mines. To a large extent, this may have been due to a pre-war British decision to purchase significant quantities of a cheaper design rather than place orders for the more expensive types of mines which the Germans opted for. Many early British mines tended not to explode on contact; and reportedly soon broke adrift from their moorings. This earned them the contempt of the Germans, and they could not be relied on as providing a means of defending strategic stretches of water or harbour approaches. Although some of the more-efficient Russian carbonit mines were acquired, it was to take some considerable time before the British were to have substantial numbers of mines that were anything like as effective or efficient as the German devices.

Lacking the benefit of effective defensive minefields made it even more

---

35 Arthur Marder, *From Dreadnought to Scapa Flow*, vol. 2, *To the Eve of Jutland 1914–1916* (Oxford, 1969), 343.

necessary to deploy ever greater numbers of small patrol vessels in this burgeoning war against U-boats and minelaying vessels. In early November there were 137 armed patrol trawlers, already or nearly commissioned, along with 37 steam yachts. This was in addition to the 246 other trawlers by then allocated to minesweeping duties, together with a couple of paddle steamers and numerous drifters that had also been taken up. Each vessel allocated for anti-submarine and patrol duties needed to be armed, and the demand for suitable guns soon exceeded the supply. It was a problem the Admiralty simply had not anticipated before the conflict. Whilst many minesweepers remained without any substantial armaments well into the conflict, those vessels on anti-submarine patrol duties used guns that were obtained from all sorts of sources, a number being reportedly acquired from Japan.

Auxiliary Patrol Weaponry

Whilst guns could be used in surface encounters, the other weaponry deployed widely on Auxiliary Patrol vessels involved in anti-submarine work was intended to be used for attacking submerged craft. The main device utilised during the first couple of years was the Modified Sweep, which was basically a long tail of armoured electrical cable to which was attached a series of heavy charges of TNT. The tail was towed under the sea and could be exploded by actual contact with the submarine or by being detonated by closing the firing circuit on board the surface ship if thought to be close enough to be effective. The sweep had many shortcomings and was cumbersome to use and probably accounted for no more than one or two U-boat sinkings during the conflict.[36]

The other weapon sometimes deployed was the lance bomb and was intended mainly for close-quarter fighting. This consisted of a conical explosive charge on the end of a wooden handle. If swung around above the head and then hurled by a strong man, then the momentum was said to allow it to cover a distance of up to 75 feet. The seven pounds of amatol in this little bomb was exploded on contact. Its effectiveness depended on the area of the submarine that it happened to strike. Despite its evident limitations it also does appear to have played a role in at least a couple of U-boat sinkings.[37]

German Minelaying in the Last Months of 1914

After laying the Tyne and Humber minefields in late August, there was a lull in German minelaying off the British coasts, and this provided a much-needed

36 Charles Domville-Fife, *The British Submarine Warfare: How the German Submarine Menace was Met and Vanquished* (Bremen, 2014), 53–54.

37 Ibid.

Photograph 2: Drifter crew on Admiralty service
Source: Edward Keble Chatterton, *The Auxiliary Patrol* (London, 1923).

opportunity for the officers and crews of the growing minesweeping fleets to gain invaluable practice and hone their clearance techniques. But in mid-October the aggressive minesweeping strategy was resumed with an initially dramatic, if rather unexpected, result for the Germans.

Around this time, the first transport ships carrying Canadian troops to Britain completed their transatlantic voyages. The Germans had anticipated the Canadian arrivals, and despatched, albeit somewhat belatedly, the old Norddeutcher Lloyd Atlantic liner, *Berlin*, to try and lay a minefield in their expected path. The *Berlin* was not only too late to accomplish this specific mission, finally setting sail on this voyage around the very time the first troopships were arriving in the Plymouth Sound, but its commander also assumed that these ships would be heading for Liverpool rather than Plymouth and by a route that would take them round the north coast of Ireland. He was wrong on both counts.[38]

The *Berlin* left Wilhelmshaven on 14 October with a cargo of 2,000 mines and after passing undetected around the north and west of Scotland arrived off Tory Island on the north coast of Ireland on the night of 22–23 October, where the former liner laid a large minefield in two legs, then ran northwards, passing between Iceland and Greenland, before putting into Trondheim in Norway with storm damage, where it was interned for the remainder of the war. On 27 October, the 5,000-ton steamship *Manchester Commerce*, en route from Liverpool to

38 Hurd, *History of the Great War*, 1:338–341.

Canada, struck one of the mines and soon sank, but it was on the following morning that the *Berlin*'s minefield claimed its most spectacular victim, the new superdreadnought HMS *Audacious*, which was steaming in the locality with the Second Battle Squadron. The *Audacious* was taken in tow, but to no avail, and eventually foundered.[39] Although the news of this spectacular loss to mines was at first supressed in Britain it was widely reported in the USA and elsewhere.[40]

Around the time of the loss of the *Audacious*, a part of the Grand Fleet, including Lord Jellicoe's flagship *Iron Duke*, had anchored in Lough Swilly on the Donegal coast, and because of the danger posed by this new minefield, whose extent was as yet undefined, they were forced to stay bottled up in this anchorage until minesweeping trawlers could commence the clearance process. Four trawlers were immediately despatched from Milford Haven for this task, and they were soon followed by two other groups. Before the end of October, six trawlers had begun sweeping an exit from Lough Swilly for the dreadnoughts. Whilst the sweeping took place, the approaches to the dreadnought's refuge were defended by the armed yacht *Lorna* and its unit of six patrol trawlers as well as a number of drifters. Trawlers, drifters, and other small craft were seemingly able to maintain their patrols in the seas outside Lough Swilly at a time when it was not safe for battleships and other large warships to venture forth.[41] In the event, no mines were found near Lough Swilly, and the dreadnoughts were finally able to leave its waters, but this ignominy was just the latest in what had proved a perplexing catalogue of unexpected occurrences in the war at sea. If one thing was clear by the end of October 1914 it was that this was already a conflict which challenged so many preconceptions of the naval strategists.

The search for the minefield continued. Additional minesweeping trawlers under the command of Sir James Donville arrived from Scapa to help locate it as weather conditions worsened with the onset of the North Atlantic winter. The full extent of the minefield was not discerned until much later in the year, somewhat further out to sea than had been anticipated, but not before the Donaldson liner *Tritonia* had foundered in waters not far from the earlier losses. Operations to clear the field lasted well into the new year.[42]

On 3 November, a German battlecruiser squadron, under the command of Admiral von Hipper, closed on the East Anglian coast and began shelling Yarmouth, also exchanging fire with inferior British forces patrolling in the area. Whilst the so-called Gorleston Raid caused only limited damage to buildings, it was used as a cover for the laying of a new area of mines, the enemy's priority here being to sink larger Royal Navy units sent to hunt down the raiding warships.

---

39 See Admiral Viscount J. Jellicoe, *The Grand Fleet, 1914–1916* (London, 1919), 149–150.

40 Ibid.

41 Hurd, *History of the Great War*, 1:339–340.

42 Ibid., 343–344.

The German raiders laid mines as they approached Smith's Knoll and then all the way down the swept channel for five or six miles, releasing yet more of these lethal weapons as they proceeded back out to sea. The Royal Navy submarine D5 was an early victim, detonating a mine near the South Cross Buoy after leaving Yarmouth to attack the raiders, but no substantial Royal Navy surface vessels were lost in the encounter, thanks in part to a report from a fishing vessel which had seen the Germans sowing their deadly cargoes.

Three fishing vessels were lost shortly after the field had been laid, but the minesweeping trawlers were soon at work, feeling their way forward in an enveloping fog. Steadily, the mines were cleared from the channel but not without the loss of the minesweeping trawler *Mary*, her skipper Stephen William Greenway, and seven other crew members: six fishermen and a Royal Navy Leading Seaman. Theirs was a peremptory form of destruction, typical of the way many minesweepers and crews were to be lost in this war, the vessel seemingly blown in two by a mine with the fore-part of the vessel sinking in no more than two minutes.[43] The *Mary* was a Fleetwood vessel, owned by J. Marr and Son and had only been hired by the Admiralty for minesweeping duties on 18 August.

Early Auxiliary Patrol Successes

The enemy's mine and U-boat offensives off the British coasts did not seem to abate during the last months of 1914 – quite the opposite in fact – and the Admiralty's appetite for requisitioning fishing vessels and other craft to counter these threats also showed no sign of being satiated. Although the requisitioned vessels of the Auxiliary Patrol were inadequately armed, they played a crucial role in an early U-boat sinking in November 1914. The U-18 had left Heligoland on 17 November with the intention of attacking warships of the Grand Fleet. It patrolled off the east coast and even at one point closed on the approaches to Scapa Flow. When just south of Hoxa Head, voyaging towards the Moray Firth, the Auxiliary Patrol vessels *Tokio* and *Dorothy Gray* spotted the unsuspecting submarine's periscope. The *Dorothy Gray*, being nearer, closed and then rammed the U-boat, bending its periscope and damaging the hull.[44]

Badly damaged, U-18 plunged downwards and attempted to escape while British warships combed the surface looking for any indication of its presence. Somewhere below them, the German crew battled with their submarine's by now recalcitrant controls, struggling for over an hour, during which time they struck the bottom on two occasions before finally breaking the surface only to

---

43 *Aberdeen Weekly Journal*, 20 November 1914. See also Gerald Toghill, *Royal Navy Trawlers, part 2: Requisitioned Trawlers* (Liskeard, 2004), 297.

44 Hurd, *History of the Great War*, 1:354–355.

be rammed by the destroyer HMS *Garry*. U-18 was finished: the crew raised a white flag and scuttled their stricken vessel; all bar one sailor was rescued by Royal Navy destroyers.[45]

The Peterhead trawler *Dorothy Gray* with its crew of fishermen was the first Auxiliary Patrol vessel to play a key role in sinking a submarine.[46] Skipper Alex Youngson and his crew were awarded £500 by the Admiralty for their 'brilliant service' while the *Tokio*'s crew were granted a further £100.[47] Their achievement illustrated that this new fishermen's navy was a force to be reckoned with in the U-boat war. Youngson (1881–1940), who had been in the first tranche of Royal Naval Trawler Reserve skippers trained at Aberdeen back in 1911, was to have a particularly active war, being awarded the MBE and mentioned in despatches on two occasions by Admiral's Jellicoe and Beatty, the former of whom boarded the *Dorothy Gray* personally to congratulate both skipper and crew. By the end of the hostilities Youngson had been promoted to the rank of Chief Skipper.[48]

Tackling the Scarborough Minefield

On 16 December, the Germans carried out their notorious raid on the north-east coast of England. Admiral Franz von Hipper's squadron crossed the North Sea and under the cover of darkness then split into two groups. The battlecruisers *Von Der Tann* and *Derflinger* headed for the Yorkshire coast and soon after dawn on that dark December day they spent 20 minutes bombarding the seaside resort of Scarborough before moving northwards to attack Whitby. Further up the coast, other battlecruisers, the *Seydlitz*, *Moltke*, and *Blucher*, attacked the Hartlepools. Altogether some 137 people, mainly innocent civilians, were killed in these bombardments, and more than 500 others were injured, an indication of the total nature of this new war. Whilst the two cruisers had been bombarding Scarborough, another warship, the light cruiser *Kolberg*, laid a minefield off the coast downwards towards Filey.[49]

The raid was not only audacious but proved fortunate for the Germans. Although pursued, seemingly trapped, and even encountered at one point by British ships, Von Hipper's squadron escaped the British warships as the weather and visibility deteriorated. The raid bore many similarities to the earlier Gorleston attack, but the Scarborough minefield laid by the *Kolberg* was much larger and soon reaped a harvest of casualties. Altogether, seven British steamships and a similar number of neutral vessels as well as two fishing trawlers

45 Ibid., 355.

46 Corbett, *Naval Operations*, 2:15.

47 Chatterton, *Auxiliary Patrol*, 37–38.

48 *Aberdeen Press and Journal*, 18 June 1940.

49 NA, ADM 186/604, 'History of British Minesweeping in the War', 14–15. See also Corbett, *Naval Operations*, 2:23–25.

and four minesweepers were destroyed before the mines were cleared. East coast shipping was disrupted for some considerable time.[50]

The formidable task of clearing the *Kolberg*'s mines began on 19 December. Minesweeping trawlers and the paddle steamer *Brighton Queen* sailed north from Grimsby, and began sweeping close to the coast between Filey and Flamborough, while the gun boat fleet sweepers, which were passing this stretch of the coast in the aftermath of the raid, were ordered to sweep a channel from Filey further northwards. In the meantime, drifters from Lowestoft were tasked with the job of warning off unsuspecting merchant vessels and with laying buoys along the channel, which was being cleared. Every difficulty faced by the crews working amongst this host of drifting mines was aggravated by the wintry North Sea weather and perpetually poor visibility. But this task could not be delayed: disruption and backlogs of sea traffic caused insurmountable difficulties at ports up and down the coast. Again, a combination of urgency and relative inexperience took a heavy toll of casualties amongst the minesweepers clearing these mines. One minesweeper, the trawler *Passing*, had a gaping hole blown in its hull. Against the odds this vessel was saved, being towed first onto Scarborough beach by the *Brighton Queen*, then later patched up and taken to Grimsby for repair before returning to Admiralty minesweeping duties for the duration of the war.[51]

It took until April 1915 to complete minesweeping operations on the field laid by the *Kolberg*, but by 26 December 1914 a narrow shipping channel had already been cleared and buoyed. This could be navigated by ships during daylight hours, but even so the losses continued into the New Year. One of the naval officers on the minesweepers, Lieutenant Hubert Basil Boothby, RNR, was blown up twice in these extremely hazardous clearance operations. First he was in command of the trawler *Orianda* when it sank on 19 December and he managed to get all his crew (except one man, deckhand James Wilson) off the doomed vessel. They were picked up by the paddle minesweeper *Brighton Queen*. Boothby soon returned to clearance duties but on 6 January 1915 he was in command of the trawler *Banyers* when it also detonated a mine.[52] Two of his crew of fishermen, Thomas Strickland and Albert Thirkettle, were lost with the ship, but the others were quite fortunate to survive. Boothby was thrown against the roof of the wheelhouse and stunned when the mine exploded. He owed his life to the trawler's cook who put his own lifebelt on his stricken captain. The survivors were landed on the Tyne by a destroyer, which had picked them up, and Boothby, who spent much of the remainder of the war on minesweeping work, was subsequently awarded a DSO and always told people he got it for losing two ships.[53]

---

50 NA, ADM 186/604, 'History of British Minesweeping in the War', 13–14.

51 Hurd, *History of the Great War*, 1:362.

52 H. B. Boothby, *Spunyarn* (London, 1929), 192–196.

53 Ibid., 199.

One positive outcome of these mine-clearance operations was the first recovery of an intact German carbonit mine, albeit without mooring chain and sinker. This was brought into Grimsby and subsequently dismantled; the detailed examination of the mechanism yielding much information invaluable to the British, who were by now endeavouring to produce mines which might match the effectiveness of enemy devices.

The Creation of the Auxiliary Patrol Structure

Even before 1914 closed, after five months of hectic requisitioning and recruiting, what was a new navy had come into being. By this time the Auxiliary Patrol, as it was by then already known, had been divided into 21 operational areas which covered every coast of the British Isles. Though there were some further modifications at various times, this operational structure was maintained until the end of hostilities.[54]

A Senior Naval Officer (SNO) was by now appointed to command each Auxiliary Patrol area. The number of patrol units at his disposal and the size of the area of responsibility was determined by the nature of the coastline, and an Admiralty judgement as to its likely attraction to minelaying vessels or submarines. Vessels and crews were still being recruited at a rapid pace and within a few months the Admiralty would have deployed 74 yachts together with 462 requisitioned trawlers and drifters across these patrol areas. These were backed up by motor boats allocated primarily for duties in coastal waters.[55] The main duties of this patrol section of the new navy was to deter and disrupt the activities of enemy submarines, and if possible destroy them, while at the same time seeking to prevent enemy minelaying by either sinking or capturing any suspicious surface craft. In all cases this called for constant and systematic patrol work on what was now a very active maritime front line.

By the end of 1914, the minesweeping forces, under the control of the Admiral Minesweeping Services (AMS), had also grown rapidly from the initial 94 vessels in the minesweeping reserve taken up during the very first days of the conflict. Around 200 trawlers were already being deployed specifically for minesweeping duties, and they were being supported by several excursion paddle steamers, which were also being drafted in. Though the paddlers lacked the range and seakeeping capacities of the steam trawlers, they were highly manoeuvrable and of shallow draft, and thus proved invaluable. Whilst some other seafarers were also drafted in to work on the paddlers, the backbone of their crews consisted of fishermen from the RNR(T).[56]

---

54 Chatterton, *Auxiliary Patrol*, 58.

55 Corbett, *Naval Operations*, 2:18–19.

56 Chatterton, *Auxiliary Patrol*, 9–10.

During 1914 some 840 German mines had been laid off the British coasts and more than 50 civilian fishing vessels and merchant craft had been lost to them. The official 1919 report on minesweeping activities estimated that one vessel was lost for every 17 mines laid in 1914 and the crude estimate was that cargoes valued at an average of £10,000 per vessel were lost with them. Given that the estimated cost of manufacturing each mine was around £150 then this seemed a very acceptable rate of return for the Germans, in the region of four times the outlay. Whilst something in the region of 300 enemy mines had been removed, later figures, obtained from the Germans after the war, revealed that around 500 were probably still effective but yet not properly located or dealt with at that time.[57]

The focus for the minesweeping forces in the southern North Sea was already that of sweeping the so-called war channel, the buoyed waters which by this time stretched northwards from the eastern end of the English Channel up to Hartlepool but was later to be extended much further northwards. Not surprisingly, given the lethal nature of mines and the lack of experience of those recruited to the task, the attrition rate in minesweepers, always high, was especially so in 1914. Even those crews that had been trained in minesweeping techniques after the RNR(T) had been formed in 1911 had only ever practised with dummy mines and got their first taste of sweeping the live versions after the outbreak of war.

The Auxiliary Patrol, this new and unanticipated navy of fishermen and fishing vessels, had by most measures experienced a baptism of fire in 1914, but had emerged as a cohesive if rather unorthodox naval fighting force, a force which would be tested almost to destruction in the following years of the Great War.

[57] NA, ADM 186/604, 'History of British Minesweeping in the War', 15.

CHAPTER THREE

The Trawler Reserve and Minesweeping: January 1915–December 1917

German Minefields and Minesweeping Operations in the North Sea and Beyond

In early spring 1915, there were about 238 minesweeping vessels in service in the seas around the British Isles. Except for some gunboats and hired paddle steamers, the majority of these were fishing vessels, at that time trawlers, their crews mainly fishermen, and their domestic sphere of operations still primarily focused on the east coast of England. Although there were about 63 minesweepers sailing out of Scottish ports, many providing protection for the Grand Fleet and other Royal Navy surface ships, and a total of around 34 craft covering the western ports or the English Channel eastwards to Portsmouth, most of the remainder were based at key eastern ports from North Shields down to Dover (Table 1).

Whilst other vessels had already been despatched for minesweeping work off the Dardanelles, the disposition of these flotillas of fishermen and hired fishing vessels reflected the main areas of German minelaying off the coasts of Britain in the first year of the war. Significant operations had recently taken place off the eastern side of the Dogger Bank and further substantial activities were soon undertaken. Presumably, such offshore minefields were primarily targeted at heavy Royal Navy warships sweeping through the North Sea or engaged in chasing German warships towards Heligoland and the like.[1] Further minelaying operations, directly involving the cruiser *Hamburg* supported by various other vessels including battle cruisers, were carried out around 18 May. Whilst these were under way, four British civilian fishing trawlers working in the North Sea – the *King Charles* of Grimsby and the *Euclid, Duke of Wellington*, and *Titania* from Hull – were captured and then sunk by German torpedo boats; their crews taken back to Germany as prisoners, presumably to keep the minelaying operations secret. Before the end of the month these trawlermen had been incarcerated in Ruhleben internment camp just outside of Berlin.[2]

1 NA, ADM 186/604, 'History of British Minesweeping in the War', 19–20.

2 *Yorkshire Post and Leeds Intelligencer*, 31 May 1915.

Table 1:
Minesweeping forces, April 1915

Base	Trawlers	Paddle steamers	Others	Totals
Orkney and Shetland	9	–	7*	16
Peterhead	8	–	–	8
Cromarty	18	–	–	18
Granton	21	–	–	21
North Shields	12	–	–	12
Humber	30	6	–	36
Lowestoft	47	–	–	47
Harwich	8	–	–	8
Nore	25	–	–	25
Dover	12	–	–	12
Portsmouth	9	–	–	9
Portland	8	–	–	8
Devonport	7	–	–	7
Milford Haven	4	–	–	4
Clyde	–	6	–	6
Totals	218	12	7	237

* Comprises a sloop and six fleet sweepers.
Source: National Archives, ADM 186/604.

Typically, these minefields were often discovered in tragic fashion, although not usually by large warships: the fishing vessels *Angelo* and *Sabrina*, owned by Hellyers of Hull, for example, probably foundered on mines laid off the Dogger Bank on 21 May, just three days after the *Hamburg*'s lethal voyage. The following month the steam trawler *Dovey* was lost on yet another such field, this one laid some 50 miles east by south of Spurn Head. At least 18 fishermen lost their lives in the *Sabrina* and *Dovey* sinkings, but the crew of the *Angelo* had better fortune, the trawler's survivors later reporting that their ship was seemingly blown right out of the water by a mine fouled in their nets. Fortunately, the ship then righted itself, just long enough for the dazed crew to get their rowing boat over the side. As they pulled away, the boiler exploded, and the stricken trawler sank stern first. Their little boat was damaged by the explosion and the fishermen stuffed their caps into a gash in its side to block the inflow of sea water. They stayed afloat, but only just; bailing and rowing almost continuously for 12 tortuous hours until they were picked up by a Norwegian steamer.[3]

The extent of the new minefields off the Humber was soon ascertained by the British and by the end of May minesweepers were reported to have already disposed of 127 mines. Nevertheless, four British steamers and five craft from non-belligerent countries were destroyed before the area was finally declared clear in the middle of July 1915.

Two more substantive minefields were laid before the Germans stopped using surface ships to deposit mines around the British Isles, and both were targeted at the Grand Fleet. On the evening of 7–8 August 1915, while a number of the Royal Navy's largest warships were anchored at the Invergordon base, the German auxiliary *Meteor* slipped unseen into the Moray Firth and sowed some 380 mines across its waters. Fortunately, the field was discovered the next morning by a trawler carrying out a routine sweep through the Firth, and although the destroyer *Lynx* was sunk, and a couple of minesweeping sloops lost their bow sections in explosions during clearance operations, none of the large warships was damaged. The minesweepers picked their way through the coastal waters and found a safe exit for the warships off the northern shores of the Moray Firth and removed more than 200 mines to create a clear passage to the southwards channel. The German mines laid in the centre of the Firth were left in place and used by the British as a defensive barrier thus reducing the sea area requiring patrols,[4] a tactic that was also used in some other North Sea waters. Meanwhile, the fleeing *Meteor* was scuttled by its master after being run to ground off the Horns Reef by Sir Reginald Tyrwhitt's Harwich fleet.

The final large enemy minefield laid by a surface craft in British coastal waters during the Great War was sown in a zigzag pattern between Sule Skerry and Cape Wrath by the raider *Moewe* on New Year's Day 1916, as it voyaged towards the Atlantic. On 6 January, these mines claimed a couple of neutral steamers and the old battleship *King Edward VII* as it voyaged from Scapa to Belfast for a refit.[5] Although 252 mines were later reported to have been discharged on this voyage, many were probably washed from their moorings by rough weather. Seventy-one mines were eventually removed to create a much-needed clear channel, but this proved a tediously incessant task for the sweepers, their crews working while exposed to the full force of the Atlantic winter.[6]

---

3 *Hull Daily Mail*, 24 May, 1915.

4 Jellicoe, *The Grand Fleet, 1914–1916*, 248.

5 Ibid., 268.

6 Archibald Hurd, *History of the Great War: The Merchant Navy*, vol. 2 (London, 1921), 260 and Dorling, *Swept Channels*, 114.

Widening the Coastal War: The Advent of German Submarine Minelaying Operations

Although enemy surface ships no longer sowed mines around the British coast, their campaign had not ended. Far from it. The Germans had merely changed direction: they had started constructing submarines that specialised in laying mines: a new and even deadlier phase of the maritime war had commenced (Photograph 3).

Whilst the sheer quantity of mines laid by surface raiders had posed many problems for the hurriedly assembled clearance crews drawn mainly from the RNR(T), the practice did at least have some apparent limitations. Surface craft by their very nature always ran the danger of being detected, and no doubt this limited the number of operations carried out. It was also difficult for surface vessels to make covert minelaying voyages to more distant regions of the British coast. Most (though of course not all) of their fields had been laid in the vicinity of the North Sea. During the surface minelaying phase, the British had been able to concentrate much of their minesweeping activities on just a few key bases. However, the introduction of the minelaying submarines ushered in a new, more sporadic, and seemingly uncertain chapter in the campaign. Although a minelaying submarine's carrying capacity was far smaller than a surface raider's, it had the potential to make many more trips as it was much more difficult to detect. Their deployment soon placed a great strain on Harwich, Lowestoft, The Nore and Dover as the busy sea lanes – the war channels that passed through their coastal waters and patrol areas – were all too well situated for regular and sometimes almost seemingly ceaseless visits by U-boats from Germany's new naval bases in Belgium, established soon after their land forces had overrun so much of that nation.

Moreover, because submarines were so much more difficult to detect, they provided a way for the Germans to sow mines in sea lanes off more distant coasts of the British Isles. Accordingly, British minesweeping forces had urgently to adapt to the new situation. The Germans had first ordered minelaying U-boats in November 1914 and developed them in great secrecy. The first type to enter surface were the UB boats and these small craft were said to be ideal for penetrating the shoals and narrows of the Thames Estuary and Dover Straits. The UC class of vessels followed and were about 40 tons larger. The first vessels could carry twelve mines. The UB boats were fitted with a couple of torpedo tubes, while the first UC boats were purely minelayers.[7]

The British remained totally unaware of minelaying U-boats until the first of these new German vessels were already in active service. The initial

[7] Julian Corbett, *History of the Great War: Based on Official Documents, Naval Operations*, vol. 3 (London, 1923), 55. See also Gibson and Prendergast, *German Submarine War*, 64–65.

Photograph 3: Scottish fishermen held in German prison camp
Source: Jon Grobler

deadly consignment was laid close to the Goodwin Lightvessel on 21 May 1915, damaging the destroyer *Mohawk*. Small quantities of mines mysteriously appeared off several different stretches of coast, including the North Foreland, Harwich, and Dover, amongst other places, and were initially put down to surface minelaying operations carried out by ostensibly neutral fishing boats.[8] After little more than a month the true cause became apparent: on 2 July, the steamer *Cottingham* struck a mysterious underwater object, and the location was dragged by minesweepers the next day. The obstruction was found and that evening a large explosion convulsed the waters.[9] When a naval diver went down on the site the following morning, he found the wreckage of a UC-2, which had sailed on only its second belligerent voyage just three days earlier with the intention of laying mines off Lowestoft. The minelaying capability of the vessel was all too evident.

Over the rest of the summer, further British losses to mines followed in swift succession, clearly caused by submarine laying activity. Deployment of the enemy's new craft was apparently rapid, the Germans later claiming to have laid 648 mines between Grimsby and Dover before the end of the 1915, with waters

8 Gibson and Prendergast, *German Submarine War*, 50.

9 Corbett, *Naval Operations*, 3:55.

off the Nore, Harwich, and Lowestoft as well as Dover bearing the brunt of such early and unwelcome attention.[10]

High levels of destruction accompanied this first phase of submarine minelaying, losses being particularly intense in the coastal waters between Lowestoft and Dover. During the last six months of the year 104 vessels were reported to have been sunk after striking mines scattered by these craft. These figures included 15 minesweepers and a further nine Admiralty patrol trawlers or drifters. Another ten civilian vessels were lost while on fishing trips. A second German minelaying submarine was lost in the autumn: UC-9 did not return after setting off on its second minelaying voyage from Zeebrugge in October 1915. The following month the body of the chief engineer, Asp Hans Neuhaus, washed ashore on the Long Sands. Fourteen crew were lost with this minelayer, which has never been found.[11]

The Germans used the captured Belgian ports as bases to great advantage for their minelaying forays. Large quantities of mines were stored at Bruges, which became the operational and maintenance centre, thanks to its canal links with Zeebrugge and Ostend. Harbour facilities at all three ports were reconstructed to accommodate substantial numbers of U-boats. Every precaution was taken to provide protection from air raids and at Ostend and Zeebrugge huge concrete shelters eventually provided safe berthing facilities for the U-boat fleets.[12]

Minelaying in the War Channels

Much of the minelaying action from the Belgian ports was now focused on the war channels created by the British, and the U-boats often disposed of their lethal cargoes in the vicinity of navigation buoys and lightvessels – not least the Sunk Lightship situated in the approaches to the Thames estuary – or off prominent headlands, such as the North Foreland in Kent. A steady stream of submarines made regular crossings from Belgium, swiftly disposed of their deadly cargoes, and then returned to base to reload. The Belgian ports were to remain much more than just a thorn in the side of British naval forces until the Germans were finally driven out quite late in the war. They were subject to British naval bombardments from the middle of 1915, and later, of course, the famous Ostend and Zeebrugge raids. Such actions involved minesweeping vessels in frontline actions as they cleared passages for attacking craft. During the bombardments of Zeebrugge and Ostend, for example, in the months of August and September 1915, the enemy shelled the vanguard of Grimsby-based

10 NA, ADM 186/604, 'History of British Minesweeping in the War', 28–29.

11 Robert M. Grant, *U-Boats Destroyed* (London, 1964), 29.

12 NA, ADM 186/604, 'History of British Minesweeping in the War', 26.

paddle steamers which cleared the way for the bombardment vessels, and on the second occasion these craft were also subject to aircraft attack when they came to the fore again to deal with mines.[13]

As 1915 drew to a close the outlook for crews of the British fishing vessels and other minesweeping forces, trying effectively to contend with this new phase of maritime warfare, looked all too gloomy. The submarine minelaying campaign seemed unrelenting on the south-east coast and during the first week of 1916 five steamships were mined off the Galloper. Meanwhile, the hard-pressed minesweepers initially destroyed 14 mines off Harwich early in the New Year, and then another ten, but not before a further ship had blown up. During February, three more steamers were then mined in the nearby war channel, and although there was a lull in minelaying activities in this area during March Harwich forces assisted Lowestoft minesweepers in clearance operations and accounted for another 11 mines.[14]

Whilst the southern North Sea coasts bore the brunt of the initial phase of the submarine minelaying campaign, the commissioning of more substantial vessels by the German Navy in 1916 intensified the campaign, increased the uncertainty, and created new and dispersed difficulties. Most notable were the new U-CII boats which not only had a greater carrying capacity – 18 mines – but also the longer range required to reach more distant waters, necessitating British forces to widen their clearance operations.

The Onslaught of the German Minelaying Flotillas

To make the most of their new capability, the Germans had created two minelaying flotillas for their operations around the British Isles. The Belgian-based vessels, which were augmented by some of the new UC class, were called the Flanders Flotilla, while the other North Sea-based UC boats, on the Elbe, became known as the High Seas Flotilla. The Elbe-based vessels, having a greater operational range, were allocated to operations in the North Sea above Flamborough Head, and also covered all coasts to the west, while the Flanders Flotilla of minelayers focused mainly on what had already become their well-known hunting ground along the southern North Sea coasts, but they did attempt to push through the Channel and beyond.[15]

Inevitably, as more vessels came into service during 1916, the Germans disseminated their submarine minelaying activities ever further around the British Isles. During April, there were also reports of the more sophisticated and

---

13 Chatterton, *Auxiliary Patrol*, 115. For an account of these bombardments, see also R. Bacon, *The Dover Patrol*, vol. 1 (London, 1919), 81–91.

14 NA, ADM 186/604, 'History of British Minesweeping in the War', 38.

15 Ibid., 33.

mobile Leon mines being strewn in the approaches to the Tyne[16] – another had been also picked up by a fishing trawler working out of the Humber three months earlier[17] – but these expensive devices remained lethal for no more than a few hours, and their experimental deployment does not seem to have been repeated. Further conventional fields were laid during the next few months in the Moray Firth and off the north-east coasts of England, including the Farne Islands area[18] and beyond, presumably with the hope of catching HMS *Marlborough* after the damage it received at Jutland had been repaired on the Tyne.[19]

Clusters of mines were also located off the mouths of the Tees and Tyne before the end of the year, but well before that time the submarine minelayers had made their presence felt much further afield, having reached as far west as Skerryvore and the coasts of southern Ireland. They had also been active along the west coast of England, with single incursions recorded in the Clyde, Liverpool Bay, and Bristol Channel approaches. On 17 August, for example, the sloop *Lavender* destroyed a German drifting mine ten miles south west of the Old Head of Kinsale and in a subsequent sweep organised from Queenstown another ten mines were cleared from the vicinity of Seven Heads. Nevertheless, the minelayers made their mark on the southern Irish coast: on 14 September, after the Harrison Liner *Counsellor* was blown up off Galley Head, patrol vessels were ordered to intercept and deflect shipping to pass the southern Irish coast some ten miles further out to sea, and soon afterwards another eight mines were found nearby.[20] A mine was suspected to have caused the loss of the SS *Kelvina*, which sank off the Helwick Lightvessel in the Bristol Channel on 2 September, and a search by Milford sweepers led to the discovery of six mines within around a week and the destruction of another 22 mines laid in an attenuated minefield that reached as far as the St Govan Lightvessel. But there were still more mines here unaccounted for, a brace of which claimed a couple of French steamers in December on the approaches to Swansea. Thirty of the 34 mines later found to have been laid in this German Bristol Channel foray were eventually destroyed.[21]

Before the end of the year there had also been mine infestations off all the principal ports along the English Channel. But perhaps the most spectacular success for the German minelayers in 1916 was the sinking of the cruiser HMS *Hampshire* with the loss of almost its entire crew, including General Kitchener who was being conveyed to Russia. The *Hampshire* blew up off Marwick Head on the west coast of the Orkneys on the evening of 6 June, having fallen victim to one of 18 mines scattered in late May by U-75, a member of the new larger class of submarine minelayers. These mines were part of a broad submarine

16 Ibid., 35.

17 NA, ADM 137/437, Humber, 31 January 1916.

18 NA, ADM 137/437, Tyne, 6 and 13 August 1916.

19 Jellicoe, *The Grand Fleet, 1914–1916*, 437.

20 NA, ADM 186/604, 'History of British Minesweeping in the War', 36.

21 Ibid.

minelaying strategy designed to catch warships from the British Grand Fleet emerging from Scapa Flow. Ironically, however, the minefield had not been sown in its intended position and 15 mines were swept up nearby.[22]

Innovation and Adaptation: Countering the German Onslaught

Adaptation, innovation, and an extension of minesweeping operations formed part of the British response to the diffusion of enemy submarine minelaying activity during 1916. A lighter form of minesweeping gear based on apparatus used by Russian and French forces proved of value. This consisted of a light wire, small kite, and a diving spar with a forward thrust that carried an explosive grapnel. This gear could be operated by one vessel and towed on each quarter at speeds of up to 12 knots. It proved useful for locating new minefields and could be deployed by smaller craft, such as the recently introduced motor launches. It also had the advantage that it could be used at night although its main drawbacks were its small spread and limited depth, but the gear was deployed with some success in several localities.[23]

One of the most important technical improvements made to minesweeping gear during the Great War was the introduction and diffusion of serrated wire sweeps. These were successfully trialled for Admiralty minesweepers by Messrs Bullivant and Company of London, wire rope manufacturers, who already made torpedo nets.[24] The fundamental modification introduced here was the incorporation of a serrated strand of wire which laid over the sweep wire, making it easier to cut through mine cables. It proved advantageous for low-speed clearance activities and made possible the use of lighter forms of gear.[25] This had the advantage of improving the minesweeping performance of steam drifters, which had less raw power to deploy than their steam trawler counterparts, and meant that more of this class of vessels could join the hard-pressed minesweeping forces.

Throughout these years, the fishermen of the RNR(T), their officers, and associated seafarers who made up much of the country's minesweeping forces, contended with all the constantly shifting currents of the minesweeping war. They were caught up in an unrelenting conflict, whether they were serving on the vessels charged with constantly sweeping the east coast war channels and port approaches or tasked with tackling the increasingly sporadic scattering of clusters of mines in waters off seemingly every coast of the British Isles.

---

22 Dorling, *Swept Channels*, 120.

23 NA, ADM 186/604, 'History of British Minesweeping in the War', 28.

24 Friedman, *Fighting the Great War at Sea*, 346.

25 NA, ADM 186/604, 'History of British Minesweeping in the War', 31.

Photograph 4:
George Frances Robinson in Auxiliary Patrol mate's uniform
Source: Author's collection

Full-scale minesweeping operations were carried out whenever daylight and weather conditions permitted, beginning at dawn and often lasting until dusk (Photograph 4).

Fishermen and Admiralty Service

The rapid expansion of minesweeping and other anti-submarine forces brought large numbers of hitherto untrained fishermen into Admiralty service. By almost every contemporary account and recollection, the assimilation of these already skilled and seasoned seafarers into military service proved to be a challenging process for all concerned. Even in an age and a society still very much used to being defined by class divisions, the gulf between working fishermen and Dartmouth- and Osborne-trained officers was enormous, and Royal Navy concepts of discipline, service, and smartness were in many ways quite alien to the fishermen. Something of the fishermen's approach to Royal Navy practice and discipline can be gleaned from surviving photographs. In the Royal Navy proper, sailors could have a full beard or be clean shaven. Fishermen in the RNR(T) could retain moustaches, which many did with great enthusiasm. Cap badges on skippers and mates were seldom straight: often a little off-centred; just the sort of feature likely to exasperate anyone who likes close conformity and order. Indeed, a lack of the sort of tidiness and order expected in the Royal

Navy, certainly in respect to dress and uniform amongst the crews, often distinguishes the fishermen recruits from Royal Naval mainstream ratings in many photographs. Fishermen generally cared little for the niceties of uniform and traditional service discipline. Yet many of those recruited, particularly the skippers and mates, who had all risen through the ranks of deckhand learner or the like, were probably some of the best seafarers of their age, used to danger, and leaders of men. Few could fault them in action once experienced. The best way of dealing with them seems to have been to avoid trying to enforce too much of the strict letter of Admiralty Regulations and to adopt some degree of a hands-off approach.

By mid-war, however, British minesweeping forces had become much more competent and efficient. Prior to the actual commencement of hostilities, not even the fishermen already enlisted in the Trawler Reserve had been given the opportunity to clear a live mine as part of their training.[26] But by mid-1916 the minesweeping service had a strong core of officers and crew who possessed a formidable level of experience when it came to dealing with live mines, and the standards of training now provided was of a much higher level than in the first months of the war.[27]

But, even though minesweeper crews were becoming more adept as they accrued experience, theirs remained one of the most hazardous of tasks. During 1916, 1,804 enemy mines were swept up in home waters[28] and every single one of these devices could send the minesweeping crews that removed them to the bottom of the sea in little more than an instant.

Although more armament was becoming available, many of the trawlers and drifters used primarily for minesweeping were still armed with little more than rifles. Wireless sets were at a premium and many vessels still communicated with shore by means of carrier pigeons carried on board. The conflict in the war channels sometimes became somewhat akin to a game of cat and mouse. Many of the submarine minelayers had at first deposited their cargoes at night, but later reports suggested that some submarines had taken up a new form of deception: lying submerged with their cargoes until daylight, sometimes waiting until the sweepers had passed them by, and only then depositing their mines in the newly swept section of the war channel to catch the first unsuspecting steamers to pass along the now ostensibly cleared seaways.[29]

---

26 Ibid., 4.

27 Archibald Hurd, *History of the Great War: The Merchant Navy*, vol. 3 (London, 1924), 227.

28 Dorling, *Swept Channels*, 338.

29 NA, ADM 186/604, 'History of British Minesweeping in the War', 42.

Reorganisation and the Intensification of Enemy Minelaying Operations

Towards the end of 1916, with the Germans threatening to wage unrestricted submarine warfare, the Admiralty reorganised and expanded its staffing with the principal aim of providing better support in both practical and developmental terms for the minesweeping forces. Hitherto, the Admiral of Minesweeping not only had full responsibility for minesweeping operations along much of the eastern seaboard north of the Dover Patrol area – itself a discrete operational district – but he had also been engaged with overseeing every dimension of minesweeping practice from the supply of equipment to the deployment of vessels and the diffusion of technological innovation. The post's workload had grown exponentially since the early months of the war and the need to spread these duties was now acknowledged. A new post was created, Director of Torpedoes and Mines, and the incumbent was charged with overseeing the development and distribution of mine protection and minesweeping gear, as well as the construction and supply of new minesweeping vessels.[30]

The pressures being placed on all ranks involved in the clearance of enemy mines from the seas did not abate. New Year 1917 heralded a further intensification in enemy minelaying operations, and this tempo continued to increase until well after the final German declaration of unrestricted submarine warfare in February. Five hundred and fifteen mines were destroyed by British minesweeping forces in the quarter ending April 1917, a figure that was almost double that of any earlier period. The onslaught continued: during 1916, a total of 195 distinct groups of German mines had been laid around the British Isles by either the Flanders or High Seas Flotillas. In 1917, no fewer than 536 separate mine clusters were deposited. It was later estimated that an average of one cargo of submarine mines was being sown off British coasts every 30 hours of the 12 months.[31]

This was a particularly perilous year for all whose business required them to frequent the sea lanes around the British Isles, not least the fishermen of the Trawler Reserve. Reports of new sightings poured into the Admiralty from far and wide. In early January, mines were found off the Hartlepools, while the forces in that area were still fully engaged in a struggle to clear difficult clusters distributed in shallow rocky waters off the Tyne and Tees during the previous month. There was no respite on the north-east coast: within a few weeks, further devices were located off Whitby and Blyth, and inclement weather continually created problems for the crews called in to clear mines.[32] Much further south, 19 mines were reported swept up off Harwich in January, but not before the loss of

---

30 Ibid., 41.

31 Ibid., 45.

32 NA, ADM 137/438, Tyne, w/e 20 January 1917.

Photograph 5: Magnus Flaws (left) from Shetland with companions *Source*: Flaws family

a Russian steamer and a fishing drifter. During February, further mine clusters were found at spots as diverse as Aberdeen, Dartmouth, Lowestoft, Flamborough, Milford Haven, Orkney, Shetland, May Island, and the mouth of the Tay.[33]

To cope with the destructive deluge in north-eastern waters, several armed trawlers normally used for anti-submarine patrol and escort duties from the Tyne were hurriedly refitted for minesweeping and turned over to the Port Minesweeping Officer,[34] a practice also followed elsewhere.

33 See Hurd, *History of the Great War*, vol. 3, chap. 9.

34 NA, ADM 137/438, Tyne, 6 January 1917.

Beyond the North Sea

Reports of mine clusters became increasingly commonplace in waters off the south and western coasts of the British Isles. By this stage, Ireland and the Western approaches proved more attractive to the larger marque of minelaying U-boats. Compared with North Sea coasts, Irish waters had previously received comparatively scant attention, at least since the *Berlin*'s cargo of mines had sunk HMS *Audacious* off Tory Island in October 1914. Hitherto, minelaying activities in southern Irish waters had been mainly confined to the vicinity of the Seven Heads Peninsula, Clonakilty Bay, and Galley Head in County Cork.[35] During March 1917, it became evident that the Germans were devoting greater attention to Irish coastal waters as new fields were found, not only in the approaches to Cork Harbour, but also off the coasts of the counties of Wexford and Waterford. The Admiralty response to these unwelcome incursions was to strengthen Irish minesweeping capabilities. Initially a new flotilla of minesweepers was sent over to Queenstown, and these were followed in April by a division of newly built paddle minesweepers.[36] These reinforcements arrived not a moment too soon as German interest in disrupting and destroying traffic through these waters did anything but diminish. During the year, 416 mines were deposited in the area by submarines.[37] The Western Approaches were a crucial strategic conduit for transatlantic supplies and the entry of the USA to the war on 6 April 1917 only emphasised their importance to the Allies.

Minelaying submarines also extended their operations in the seaways off the west coast of Scotland during the first months of 1917. Mines were found off Mull, Stornoway, Coll, Skye and Harris, the Clyde Approaches, and also Lough Swilly on the Irish north-western coast. In the spring of that year enemy minelaying submarines were reported to be frequenting the Little Minch in the Hebrides and various tracts of water in the vicinity that fleet destroyers and auxiliary vessels were known to use. This was initially countered by moving not the usual Auxiliary Patrol vessels to the area, but the old gunboat flotilla which had been formerly guarding the Grand Fleet and was now displaced from those duties by new and fast twin-screw sweepers. The flotilla was ultimately based at Oban and placed under the full control of the Director of Minesweeping Operations but had a wide remit which could take them out of that area if required. In view of this, a local sweeping force was created using six motor drifters. These were commissioned because of the shortage of steam trawlers and drifters but proved less than successful, hampered by their slow speed and the lack of efficient motormen, and were eventually replaced by steam vessels when these became available.[38]

---

35 Edward Keble Chatterton, *Danger Zone* (London, 1934), 199.

36 Dorling, *Swept Channels*, 125.

37 Ibid., 129.

38 NA, ADM 186/604, 'History of British Minesweeping in the War', 54.

The approaches to Stornoway and the surrounding area were also visited by the German craft. Several mines were found in these waters during February and March, and in the latter part of April a moored mine was found by Stornoway sweepers off the Shiant Isles, followed in May and early June by another ten devices in the Stornoway Approaches. In July, a field off the Butt of Lewis was discovered and eight of these were removed by the Stornoway sweepers. A group of four more laid off Loch Ewe were also detected and destroyed. Yet another 12 were laid in Butt Bay, erroneously believed by the Germans to be a regular anchorage for passing traffic. They did no damage and were not located until much later in the war.[39]

Many other mines were discovered and eliminated in the wider north-west region but only after the loss of four minesweepers, a destroyer, and four steamships, including the armed merchant cruiser *Laurentic*, which struck two mines off Lough Swilly on the bitterly cold and stormy evening of 23 January 1917. She sank with the loss of 349 people and specie to the value of £5 million while outward bound from Liverpool to Halifax, Nova Scotia. Many of those who lost their lives died from exposure in open boats.[40]

On 15 February, the White Star liner *Celtic* was damaged by a mine in the Irish Sea, but then towed into Liverpool for repairs. The steamer *Kelvinhead* struck another mine and foundered off the Mersey Bar on 27 March, and on 8 April, the transatlantic liner *Lapland* was damaged in the same vicinity but also managed to make Liverpool and safety. The following evening yet another liner, this time the *New York*, also struck a mine. Although a blizzard was raging and a nasty sea running, all the passengers were safely transferred into lifeboats and picked up by vessels which hurried to the rescue. Those brought safely ashore included Rear-Admiral William Sims of the United States Navy.[41] Three days earlier the USA had entered the war, and Sims assumed command over all the US Naval forces operating in Europe. Despite being holed under the waterline, the *New York* also made it into dry dock.

The narrow nature of the deep-water ship channel leading to Liverpool meant effective mine clearance was particularly difficult. All too often mines that had been swept up yet had sunk before exploding were thought to be uncomfortably close to passing deep draught shipping. As Liverpool was to be a principal point of entry for US troops, the port's minesweeping capabilities were increased, but this coincided with a long lull in enemy minelaying visits to the Mersey approaches. Unfortunately, this state of affairs was not permanent: the Germans resumed their interest before the end of the year and in the early hours of 28 December the Mersey pilot cutter *Alfred H. Read* struck a mine on the bar station and sank in minutes. This was probably the worst disaster that has ever

---

39 Ibid.

40 Dorling, *Swept Channels*, 124.

41 Gibson and Prendergast, *German Submarine War*, 155.

befallen the Liverpool Pilot Service and of the 41 pilots and crew on board the stricken vessel only two people survived.[42]

The traffic in shipping lanes around the coasts of the British Isles, which the enemy continued to target and the minesweepers attempted to protect, remained immense: 30,000 vessels were reported to have passed Lowestoft in the war channel during 1917 and on one occasion, when the port of Liverpool was reopened after being closed for 48 hours while minesweepers destroyed a number of newly placed mines, some 114 vessels, representing a quarter of a million tons, including two of the world's largest ships, sailed in and out of the Mersey on one tide.[43]

Fishermen and the Response to the 1917 Minelaying Onslaught

There was by now yet another new dimension to the warfare. The larger class UC boats also carried torpedoes. It appeared that when an area was clear of outlying shoals these craft often made it their practice to lie in a seawards position after sowing their mines in the shipping channel and then used their torpedoes on vessels diverted into their path when the mines were discovered. Such tactics had to be countered and in the Dover Patrol area the practice of prohibiting shipping entering a one-and-a-half-mile circle around the first mine found was introduced.[44]

There was only one east coast area where the enemy minelaying onslaught showed much sign of abatement in 1917, and that was in the waters off the Humber – an area that had been earlier heavily mined offshore.[45] Little in the way of relief was afforded other mine clearance forces on the east coast of England in the months following the declaration of unrestricted submarine warfare and the minesweeping forces working off Harwich were kept particularly busy. The Germans were intent on blocking the approaches to this key naval base as well as disrupting the passage of mercantile traffic along the adjacent war channel. Between the beginning of the year and the end of May 1917, the local auxiliary forces responsible for mine clearance destroyed a staggering total of 288 mines in the area: 84 of these alone were eliminated in the month of March. In April, the numbers accounted for reached 108, many of which were found in five separate fields distributed along the War Channel.[46] At least two minesweepers were lost in operations, but no larger surface warships or merchant steamers fell victim despite the intensity of the enemy depredations. Drifters allocated to this area

42 Stephen McGreal, *Liverpool in the Great War* (Barnsley, 2014), 128.

43 NA, ADM 186/604, 'History of British Minesweeping in the War', 45.

44 Bacon, *Dover Patrol*, 1:118.

45 NA, ADM 186/604, 'History of British Minesweeping in the War', 59.

46 Ibid., 56.

were also equipped with the new lighter serrated wire minesweeping gear and could shoulder something of the increased burden falling on the other Harwich minesweeping vessels over the remainder of the year.[47]

At Harwich, as elsewhere, English and Scottish fishermen continued to form the bulk of the clearance crews. Whilst minesweeping almost anywhere off the British Isles was by this time an almost unrelenting task wherever it was undertaken, it seemed to be particularly ceaseless on this stretch of coast. As fast as mines were cleared from the nearby channels they were laid again by Flanders Flotilla submarines making the relatively short crossing from their Belgium bases. Channels had to be swept clear on every day that weather conditions permitted. The minesweeping vessels and their crews usually spent four days in a row on duty, often starting at daybreak and sometimes continuing for many hours despite the most hazardous of conditions. The most appropriate time to sweep the channels was at highwater but limiting operations to such times was often a luxury that could be ill-afforded, given the wartime imperative to keep vital cargoes moving up and down the sea lanes. The risks posed by mines at low water, particularly to deep-draughted sweeping vessels such as trawlers, were especially high, but when shipping traffic was heavy, and the tides were at awkward times, or when the weather had previously caused disruptions, the officer commanding the local war channel forces was faced with choosing one of two extremely difficult courses of action. He could either hold up all steamer traffic, delaying the movement of vital seaborne supplies, and wait for high water when the channel could be swept with marginally less hazard, or look to minimise shipping delays by risking vessels and men on an often dangerous low-water sweep.[48]

Something of what fishermen from the Trawler Reserve experienced along this stretch of coast when on frontline minesweeping duties during this period can be gleaned from the diary of George Robinson, in peacetime a mate on the Hull boxing fleet trawlers, who maintained his journal throughout much of 1916 and 1917. Robinson had joined the Trawler Reserve in April 1915 and transferred to the hired paddle steamer HMS *Bourne* – formerly the excursion vessel *Bournemouth Belle* – on 21 March 1917, having previously seen service on HMS *Seamew* and several other minesweeping trawlers.[49] Over the rest of the year *Bourne* first worked out of Sheerness and then Harwich, sweeping the war channel and the Thames Approaches. Whilst many, though certainly not all, of his diary entries are somewhat laconic in nature, they illustrate something of the lot of minesweeping crews at this time: long spells of drudgery, day after day of sweeping dull grey seas without any result in all sorts of weather,

---

47 Ibid., 56–57.

48 Ibid., 76.

49 NA, BT 377/7, Service Record, SA 1505, George Robinson and Diary of G. Robinson, 2nd Hand, HMT *Edward VII* from the day of December 10th when joined up for sweeping enemy mines in the North Sea in the year 1915 & after (original in possession of author). Hereafter Robinson, 'Minesweeping Diary'.

operations often starting soon after sunrise. The work could be frustrating and was always hazardous. The sweep wires often parted and the need for the vessels to manoeuvre closely together in all sorts of seas when commencing towing operations meant the danger of collision was all too apparent; indeed, after mines, collision was the second most common cause of armed auxiliary trawler losses. Suddenly, weeks of drudgery gave way to spells of intense, almost frenetic activity and sometimes high tension when for day after day mines were found in the seaway. Then there was the occasional narrow escape or encounter with enemy craft. The diary shows that between the last day of March 1917 and the end of the year the crew of HMS *Bourne* was involved in the destruction of at least 74 enemy as well as a small number of drifting British mines, any of which could destroy both vessel and crew.[50]

Whilst vessels like the *Bourne* swept in tandem, its sweeping partners changed regularly during the year, the vessel linking up at different times with a variety of different auxiliary craft, sometimes returning to tying up by the buoys in port on an evening, but on other occasions lying-to during the hours of darkness off various parts of the coast. Although nominally minesweeping crews worked for four-day spells, followed by a break in harbour for refuelling and the like, it is clear from his diary that there were occasions when vessel and crew were either continuously at sea, or ready to sail for many more days on end, and also spells when crew remained on continuous guard because of air raid warnings which became an all too regular feature of the war on the south-eastern coast.[51]

Throughout these months, vessels and people well known to Robinson and no doubt the rest of the crew were lost. On 29 October 1917, for example, he notes that the minesweeping trawler *Strymon* was sunk by a mine – later shown to have been laid by the UC-11. Ten crew were lost including the skipper who was an old friend. Robinson had been talking with him only the evening before and, as he starkly remarked in his diary, he 'wished me good night and good luck and next day dead. Never know when my time is coming'.[52] A similar fate seemed sometimes most likely, but both the *Bourne* and Robinson survived, although not without incident: on 23 August both had a lucky escape when a mine was found entangled in their sweep during hauling, and, beforehand, on 20 July, he confided to his diary that he had been as 'near death as I had ever been' when the lookout failed to see a floating German mine which they managed to clear, but only after an anxious few seconds. On the same day, the minesweeper, *Queen of the North* – in peacetime a Blackpool paddle steamer – was less fortunate, being blown up with the loss of 29 crew.[53] Minesweeping was essentially a defensive reactive form of maritime warfare but there were, it is clear from the

50 Robinson, 'Minesweeping Diary', 21 March–31 December 1917.

51 Dorling, *Swept Channels*, 264.

52 Robinson, 'Minesweeping Diary', 29 October 1917.

53 Ibid., 20 July 1917.

diary, occasional chances to carry the war back at the enemy, as on 7 July, when the crew opened fire on German aircraft as they headed for Margate, and then renewed their fusillade when they flew overhead once more after turning for home.[54] On another occasion, the *Bourne* attacked a U-boat which had surfaced nearby and attempted a ramming, though without success.[55]

The *Bourne* did not escape unscathed from the Great War. She was badly damaged on the evening of 16 December 1917, while at anchor in Yarmouth Roads. With the wind blowing hard the paddle steamer *Waverley* drove on to the ship. The *Bourne* was nearly lost in the collision but managed to get clear of the other vessel. Though in a serious state, the ship could not be immediately repaired – such was the pressure on east coast marine engineering facilities – and remained on duty for the rest of December before finally being despatched up to Grimsby for refitting.[56] Both Robinson and the *Bourne* remained in Admiralty service throughout the last year of the war and were demobilised and decommissioned sometime after the Armistice.

Improving Performance and Turning the Tide

The minesweeping crews clearing this line of coast were clearly operating at full stretch for long periods of time, but an examination of the statistics shows that their efforts met with substantial success. German records later revealed that 2,450 mines had been laid off the Harwich and Lowestoft coasts during 1916 and 1917, and of these 2,084 were cleared up by minesweepers. Although 40 per cent more enemy mines were deposited in these areas during 1917 compared with the previous year, there was a fall in the number of vessels that they sank. Whilst 28 merchant ships had been mined in the Harwich area in 1916, for example, only seven succumbed in the same fashion the following year. The intensity of the German minelaying effort on this stretch of coast can be gauged from a survey of figures relating to the section of the war channel that became the complete responsibility of Harwich-based sweepers from March 1917. During the full year, the Germans deposited 680 mines in what was only a 19-mile fairway with short connecting channels. A post-war analysis of enemy activities here reveals that 468 of these were sown within two miles of the Shipwash Shoal. Certainly, the safety of the immense flow of vessels through the Harwich war channel was considered, in an official post-war report, to be 'entirely due to the efficiency of the organisation of local minesweepers'.[57] By the very nature of their operations in and around this very heavily mined stretch of the east coast war channels the Harwich sweepers

54 Ibid., 7 July 1917.

55 Ibid., 12 November 1917.

56 Ibid., 16 December 1917.

57 NA, ADM 186/604, 'History of British Minesweeping in the War', 58.

were involved in rescuing passengers and crews from sinking vessels. From the establishment of the Harwich base up to the end of December 1917, 1,065 men women, and children had been plucked from the seas by the port's minesweepers.[58]

Further innovation and adaptation also assisted in the war against mines during 1917. Motor launches, with their shallow draft, were increasingly used to locate newly laid mines at low water and this helped reduce losses of the deeper-drafted minesweeping trawlers. There were also indications that vessels deploying the new hydrophone listening devices could prove useful in this struggle: the noise of one minelaying submarine, for example, was picked up by hydrophones off the Isle of Wight in 1917. The next day minesweeping drifters were called in to investigate, locate, and destroy the deadly consignment.[59] But another very effective development in the fight to keep ships safe from mines was the introduction of the paravane. This apparatus, which consisted of cables, cutting gear, and kites, was towed from both bows of a ship. Designed by Lieutenant C. D. Burney, it was introduced for warships in 1916 and the following year a version known as the otter began to be fitted to merchant ships by Messrs Vickers.[60] By September 1917, 294 merchant ships had been fitted with the otter gear and by the Armistice in November 1918 the number had risen to about 3,000.[61] Many years later, after the Second World War, Burney used his paravane experience to design the first commercially successful stern trawling gear used on fishing vessels.

The convoy system was also finally introduced in 1917 and although probably of considerable significance in changing the course of the maritime war on commerce it was perhaps more effective in countering U-boat torpedo and surface attacks on merchant vessels than on curtailing minelaying activities. The areas the shepherded vessels passed through had still to be kept swept and the seaways followed by the convoys remained important targets for the minelayers.

The prospect of an abrupt death remained an ever-present reality for every minesweeping crew. Typical that year was the loss of the trawler *Strathrannoch*, which struck a mine while sweeping off St Abb's Head and was blown completely to pieces. Not a survivor was found, not even a spar: nothing was recovered except a lifebelt.[62] However, for the Germans submarines, the mine proved a double-edged weapon: a substantial number were destroyed by their own mines, including the UC-32 which had crossed the North Sea in February 1917 with orders to mine shipping routes and port approaches off the north-east coast of England. She blew up on the surface while minelaying off Sunderland harbour with 14 devices still in the tubes. Only three out of the crew of 25 submariners survived.[63]

---

58 E. F. Knight, *The Harwich Naval Forces: Their Part in the Great War* (London, n.d.), 215.

59 Hurd, *History of the Great War*, 3:236.

60 L. Cope Cornford, *The Paravane Adventure* (London, 1919), 231.

61 Ibid., 244.

62 Hurd, *History of the Great War*, 3:234.

63 Paul Kemp, *U-Boats Destroyed* (London, 1997), 25.

During 1917, the Admiralty had increased its minesweeper capability. At the end of 1916 the Admiralty minesweeping forces consisted of 93 fast sweepers, largely paddle steamers, sloops, and gun boats, and 523 slow steamers, mainly trawlers, drifters, and motor launches. But a year later, as Table 2 shows, despite further losses in action, these numbers had reached 112 and 643 respectively. The increase in the numbers of slow sweepers was made possible by the further requisitioning of fishing vessels and in part by the temporary transfer of vessels normally used on other patrol duties, but by this time new classes of trawlers, drifters, and paddle steamers, specially built by the Admiralty, to standard designs, were coming into service and more motor launches were being deployed for inshore mine location work.

Table 2:
Minesweeper capability, 1917–1918

	January 1917	1 July 1917	1 January 1918
Fast sweepers (including sloops, gunboats, training ship sweepers, and paddle minesweepers	93	112	122
Slow sweepers (including trawlers, drifters, and motor launches)	523	643	631

Source: National Archives, ADM 186/604.

The picture off other United Kingdom coasts reflected in many ways the drama that was played out off East Anglia and in the Thames Approaches. Although more than double the number of mines were swept up in 1917 around the British Isles compared with the previous year, and the area mined by submarines more than trebled in extent, overall losses of merchant ships fell by 20 per cent. In terms of the minelaying offensive, the increased efforts of the Germans were accompanied by diminishing marginal returns and this trend became more evident as the year progressed. Although other innovations played their part, much of the responsibility for the improved British performance against minelaying submarines in 1917 must be attributed to the improved effectiveness of the minesweeping forces, the backbone of which continued to be formed by fishermen of the Trawler Reserve.

Chapter Four

Offensive Actions

Auxiliary Patrol Operations in 1915

Whilst the kernel of the Trawler Reserve had been created prior to the war, born of the perceived need to develop a defensive force capable of dealing with the threat of offensive enemy minelaying, the idea of extending its operations into other, often more overtly aggressive spheres of action had evidently not been anticipated, and certainly not planned for, to any considerable degree. But, as we have seen in Chapter 2, within a few weeks of the opening of hostilities, large numbers of additional fishing vessels and fishermen were quickly recruited by the Admiralty, not only for defensive minesweeping duties but also for a range of other, often more directly belligerent activities. By early 1915, less than six months into the war, the nature and range of these additional frontline operations, of what was for all intents and purposes a new and improvised navy,[1] were becoming somewhat more clearly defined.

The opening months of the war had seen quite extensive activities by enemy surface minelaying vessels and submarines off the British coast. Such operations were also thought at the time to provide a means of allowing spies to make covert landings in the country. General instructions issued to Auxiliary Patrol units stationed around the British coasts in early 1915 were specifically focused on dealing with these threats, not only by trying to prevent such incursions occurring in the first place but also by ensuring the harassment, capture, or destruction of the enemy vessels involved.[2]

To carry out such duties, each district took into account the local geography and regional strategic priorities when formulating operational plans.[3] Vessels operating in the Orkney and Shetland Archipelago, for example, not only had to anticipate enemy attacks in and around the hugely important Grand Fleet base at Scapa Flow, but were also required to sink or harass U-boats voyaging westwards by way of the top of Scotland, most notably through the Fair Island Gap. By spring 1915, many Orkney duties were being carried out by a fleet

---

1 Chatterton, *Auxiliary Patrol*, 1–3.

2 NA, ADM 137/1049, Auxiliary Patrol Organisation, April 1915.

3 Chatterton, *Auxiliary Patrol*, 58.

of eight yachts and 72 armed trawlers, supported by a few drifters. These vessels were divided amongst three patrol areas based on Kirkwall, Stromness, and Longthorpe, known respectively as the Northern, Western, and Southern Patrols. Further north, the Shetland Patrol, based at Lerwick, was somewhat smaller in strength, consisting of three yachts and 18 trawlers, although its patrol area was quite extensive. The Auxiliary Patrol units deployed in these northern waters not only had to scour the open seas for enemy activities but also to search the abundant inlets, channels, and rocky coasts of these archipelagos. These tasks often involved the examination of all manner of vessels passing through the area.[4]

Similar variations in the strength of vessel allocations and duties were found elsewhere. There was no major surface warship base in north-east England and this was undoubtedly the reason why the Tyne Auxiliary Patrol had far fewer vessels stationed in its area than either the Granton or Orkney bases, which covered Rosyth and Scapa Flow respectively, even though an extensive length of coastal waters, stretching from north Northumberland down to Scarborough, frequented by heavy commercial steamer traffic, fell within its remit. The twelve patrol trawlers then stationed at the base were divided into three groups of four vessels, known respectively as groups a, b, and c. Unless required for special duties, these groups were responsible for patrolling the coast of north-east England, which was further divided into two operational sections, the first of which stretched from Berwick to Blyth (Area 1) and the second from Blyth down to Scarborough (Area 2). Operational practice required each group to spend six days at sea followed by three days back in harbour, and while out on patrol each group would spend three days working Area 1 followed by three days examining the coasts in Area 2 before returning to base.[5]

The group of four patrol trawlers working in each of these sections was further split into pairs to try and ensure that there were always two craft off the coasts of the northern and southern sections of these two main patrol areas. In Patrol Area 1, for example, two vessels would be expected to maintain a presence northward of the Farne Islands, while the other two would focus their attention on the coastal waters from the Farnes down to Blyth. The group of vessels working Patrol Area 2 would similarly divide their activities between the areas to the north and south of Sunderland. Although these patrol groups were expected to concentrate on covering the waters within a few miles of the coast they were also required to make periodic searches to the seaward; these forays got under way at noon on prearranged dates, the armed trawlers involved steaming out to a predetermined position that might be as far as 50 miles from the coast.[6] The Tyne vessels sometimes found it difficult to cover such a wide

4 NA, ADM 137/1049, Auxiliary Patrol Organisation, April 1915.

5 Ibid.

6 Ibid., 25 April 1915.

range of operational duties, as there was little or no spare capacity to fall back on at the base and this was the cause of frequent complaints from the SNO Tyne to the Admiralty at this time.[7]

The patrol areas whose remit covered various stretches of the English Channel also took similar account of both geography and the limited resources in terms of men and ships that the Admiralty could allocate. The Newhaven base, which had responsibility for the coasts and busy seaward areas between Dungeness in the east and the Royal Sovereign Lightvessel to the West, maintained three divisions of three trawlers, each division being under the command of a Lieutenant RNR and two of these divisions were expected to be constantly at sea unless driven to seek shelter due to stress of weather. Patrol periods there then lasted for twelve days followed by six days in harbour.[8]

Further westwards, the Poole Auxiliary Patrol base covered the broadening sea areas of the English Channel from Anvil Point near Swanage to Dartmouth in Devon, then southwards towards the French Coast. By the early months of 1915, operational activities here were similarly divided into two sections: the Western Patrol boundaries were defined as being between a line joining Dartmouth and Les Sept Iles off the north coast of Brittany to another line passing between Portland Bill and a group of sea-washed rocks called the Casquets, some eight miles north-west of Alderney in the Channel Islands, and the Eastern Patrol area stretched from this latter imaginary line to another joining Anvil Point to Cape la Hague at the tip of the Cotentin Peninsular in Normandy. The patrols from Poole base were carried out by four units. The intention here was to try and maintain a patrol of six trawlers (two units) in the Western area while another unit covered the eastern area; the other or spare unit being on stand-off for the week back at base in Poole harbour.[9]

Similar arrangements, with comparable regional variations, were in place in all 21 patrol areas that had been created, and, in theory, ensured that all the coastal waters around the British Isles were being watched and guarded by vessels of the Auxiliary Patrol, most crewed largely by fishermen from the Trawler Reserve. However, coverage was not only often quite thinly spread in areas away from the strategically important Scottish warship bases, but often quite challenging to maintain. The limited numbers of armed trawlers and the like then available made it difficult to replace vessels requiring refitting or repair and the need to respond to wartime incidents or participate in special operations stretched Auxiliary Patrol resources to the very limit.[10]

The Straits of Dover were one of those areas overseen in a somewhat different manner. The Dover Patrol was a distinct Royal Navy unit. Its area of

---

7 Ibid., 25 April 1915.

8 NA, ADM 137/436, Auxiliary Patrol Tyne, 23 June 1916.

9 NA, ADM 137/1049, Auxiliary Patrol Organisation, 25 April 1915.

10 NA, ADM 137/436, Tyne, 23 January 1916 and 31 March 1916.

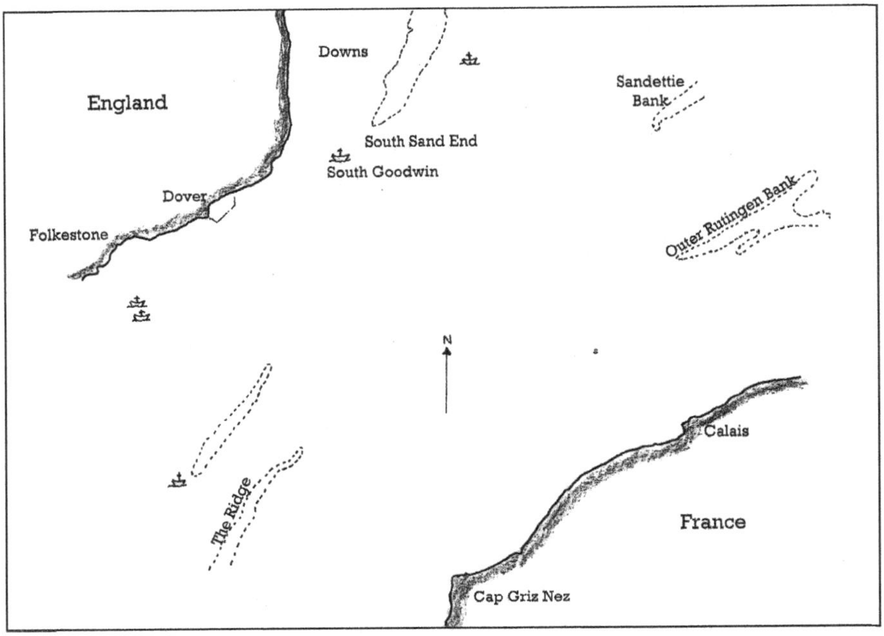

Figure 5: Outline map of the Straits of Dover
Source: Ken Knox

operations encompassed the Dover Straits and adjacent North Sea waters, and over the course of the war it deployed a range of sometimes substantial surface warships as well as aircraft, paddle minesweepers, armed trawlers, and drifters (Figure 5). Its remit covered a wide range of activities including minesweeping, minelaying, anti-submarine and escort duties, constructing and maintaining barrages, and bombarding German bases on the Belgium coast. Fishermen were employed in many of these operations, and fishing vessels in service here combined minesweeping, escort, and anti-submarine duties.[11]

Armed Fishing Vessels on Patrol and in Action during 1915

Though the ranks of the trawlers were increasingly reinforced by drifters, many given specific patrol duties, particularly in terms of the newly created east coast war channels, their armaments left much to be desired. Whilst virtually every trawler allocated for these Auxiliary Patrol duties was armed, many were still fitted with 3-pounder guns or at best 6-pounders, while the enemy U-boats

[11] Bacon, *Dover Patrol*, 1:115.

usually possessed much larger calibre weapons. This meant the German craft had the potential both to outrun and outrange the Auxiliary Patrol vessels. Contemporaneously, there remained little in the way of effective additional kit available to armed trawlers with which to locate and attack submerged U-boats: in early 1915, and for a long time afterwards, many of these vessels deployed the modified sweep outlined in Chapter 2, which was dragged through the water while the hydrophone – a form of electronic underwater listening equipment – and the depth charge were still in their early developmental stages.

During 1915, there were numerous encounters between armed trawlers and U-boats in which shots were exchanged, but these usually ended with the U-boat outpacing its pursuers and either submerging or disappearing over the horizon. This was particularly a problem for the Trawler Reserve crews charged trying to prevent U-boats passing through the Fair Isle Gap, and in May and June 1915 there were a few somewhat similar encounters, typical of which was the one involving the Armed Trawler *Ontario*, which sighted a large U-boat bearing 59.30 North, 1.53 East, and closed at full speed. At 3,000 yards the trawler's gun opened up, firing 21 rounds in all, a couple of which were reported to have hit the enemy's hull. But this was to no avail: the U-boat used its much superior speed to rapidly outrange the trawler on the surface. To rub salt into the wound, the *Ontario* could not relay details of the encounter back to base until it met with a wireless-equipped trawler.[12]

Indeed, inefficient communications also constrained the effectiveness of most Auxiliary Patrol operations in the early years of the war. Although speed and the rapid relaying of intelligence and orders were clearly important when hunting and tracking an enemy submarine's course through various sightings, few armed trawlers and drifters were equipped with wireless. Moreover, the Admiralty long remained reluctant to grant requests from Auxiliary Patrol SNOs for such equipment to be fitted, partly on the grounds that the increased radio traffic might interfere with what were perceived to be more important naval communications. Accordingly, it was difficult to expedite orders swiftly. In April 1915, for example, as the SNO Granton pointed out, an instruction for all Auxiliary Patrol vessels to return to base took many hours just to be taken to outlying craft in his area.[13]

Owing to the dearth of on-board wireless sets, many messages were passed to and fro by means of coastal signal stations. On the east coast of England, for example, 13 war signal stations were set up between Newton by the Sea in Northumberland and Cromer in Norfolk. Unless they were in company with a yacht fitted with wireless the armed trawlers and drifters out on patrol were normally expected to make contact by means of Morse or semaphore with one or other of these shore-based signal stations at sunrise, noon, and sunset in order

[12] NA, ADM 137/138, Orkney and Shetland, 29 May 1915.

[13] NA, ADM 137/201, Auxiliary Patrol Granton, w/e 26 April 1915.

to report and receive relevant information.[14] When ordered to patrol well out into the North Sea and away from the coast, armed trawlers without wirelesses usually carried a number of pigeons on board which were to be released periodically, or when enemy submarines were sighted. Many people in the east coast ports kept carrier pigeons and at Grimsby, on the suggestion of Commander Boothby, a special loft was established. Local owners lent their birds and every section was able to carry enough pigeons to send reports to shore about anything of interest.[15] Similar arrangements were in force on most sections of the coast. In the Poole Auxiliary Patrol area, a body of coast watchers was set up to facilitate signalling through local coastguard stations and included not only a chief petty officer but also local fisherfolk and boy scouts.[16]

The stakes and tempo of the war at sea were, of course, raised markedly when Germany opened its first U-boat campaign against commerce in February 1915. Attacks on merchant steamers increased immediately. These actions piled greater pressure on the already stretched men and ships of the Auxiliary Patrol who had both to respond to the rising number of U-boat sightings and go to the aid of vessels and crews in distress. Attacks on unarmed fishing vessels working out on North Sea grounds also began in earnest from early May,[17] and the necessity of sending armed vessels to guard trawlers and drifters working many miles out to sea could often only be achieved in many of the North Sea Auxiliary Patrol areas by cutting back on the scheduled patrol duties. Stretches of the coast were sometimes left unguarded.

Before long, there was yet a further substantial dimension of enemy submarine activity for the Auxiliary Patrol forces to contend with. The more random and geographically diverse nature of minelaying which followed the deployment of new minelaying submarines from June 1915 not only placed greater pressure on the Admiralty's mine clearance forces it also had profound implications for those Auxiliary Patrol vessels and crews charged with the task of harassing and trying to eliminate submarines and minelaying craft of all kinds.

Although the Admiralty faced some formidable logistical constraints in trying to keep pace with this rapidly expanding coastal war, it could be argued that their response, especially in the way that the Auxiliary Patrol was deployed, did contain some elements that were quite innovative. From the very earliest months of the war the Admiralty had looked to find ways of overcoming the advantages that U-boats possessed over armed fishing vessels in terms of both speed and armaments and in some ways the manner in which Auxiliary Patrol vessels were instructed to operate anticipated the later and rather specific

[14] NA, ADM 137/1049, Auxiliary Patrol Organisation, 25 April 1915.

[15] Boothby, *Spunyarn*, 197.

[16] NA, ADM 137/1049, Auxiliary Patrol Organisation, 25 April 1915.

[17] R. Robinson and I. Hart, *Viola: The Life and Times of a Hull Steam Trawler* (London, 2014), 85.

Photograph 6: HMT *Ferriby* on Admiralty service
Source: Charles Dinsdale

operational role of the so-called Q ships. In the first month of the war Jellicoe had ordered that armed patrol trawlers – unlike minesweepers – should retain their civilian company funnel markings and not be painted grey like other warships. Patrol trawlers thus retained their fishing numbers and many markings, and the pretext for this was in part to try and convince an enemy U-boat that they were unarmed fishing craft. Instructions issued to the vessels operating from the Tyne base, for example, were explicit on this front, stating that armed patrol craft must not only look like fishing vessels but must act as such if they ever hope to get in a shot at a submarine, or the opportunity to try and ram it, as it was believed by the Admiralty that enemy submarines, stalking by periscope, only came to the surface and attacked after they were thoroughly satisfied they were tracking an unarmed vessel. Patrol trawlers were told to stop occasionally, as if hauling in a trawl, and that no one should be on deck with either binoculars or telescope. Lookout duties were to be carried on from the wheelhouse. If a submarine did appear to be taken in and approach, then the armed trawler crew was instructed not to make an immediate rush for their gun but to pretend to get their open boat ready to put over the side. Such subterfuge was to be abandoned when the enemy had closed. Then firing was to commence as soon as possible and certainly before the submarine had the chance to submerge or get out of range.[18]

18 NA, ADM 137/1049, Auxiliary Patrol Organisation, 25 April 1915.

This basic tactic does not appear of itself to have been particularly successful and there seems little evidence in the surviving Admiralty reports from the first twelve months or so of the war of many armed trawlers or drifters being mistaken for unarmed fishing vessels while carrying out orthodox patrol work. This was perhaps not surprising as information gleaned during an 'informal' conversation between a Royal Navy commander and the captured German Captain of U-40 revealed that U-boats often kept a close watch on potential targets before attacking, and that they could often distinguish groups of patrol craft from fishing vessels because the former were often moving too fast to be trawling, or else did not have warps down or other indicators, such as baskets on the mast head in place, that suggested they were fishing. Their guns were often also too prominently positioned.[19]

However, more elaborate subterfuge evolved and clearly did work on several occasions which is how the U-40 commander came to be a prisoner of the British in the first place. The trawler *Taranaki*, for example, was one of several vessels assigned for special duties in the late spring and summer of 1915, being sent out in the guise of a civilian fishing vessel to sail in an area known to be frequented by U-boats. *Taranaki* was to act as a lure and was not alone but connected by a combined towing and telephone cable to the submerged submarine C24. On the morning of 28 June, the trap was sprung some 50 miles south-east of Aberdeen when the U-40 surfaced near the *Taranaki* and ordered the crew to abandon ship. Whilst the U-boat was concentrating on shelling the *Taranaki*, the submerged C24 prepared to attack. But all did not go according to plan as it proved impossible to slip the connecting cable. There was a real chance that the submarine would be dragged down by the stricken trawler or at least spotted by the submarine if ballast tanks were blown to correct the trim. Using his dive planes instead to correct his trim, the C24's commander, Lieutenant Frederick Taylor, regained some degree of manoeuvrability and got a torpedo on target which exploded under the German's conning tower. The commander of the U-40 was one of only three survivors and afterwards admitted that he was completely deceived by the *Taranaki* and had been watching the vessel all morning before attacking.[20] The *Taranaki*, though damaged by the shelling, was also saved.[21]

A few weeks later there was a second success for this stratagem when the trawler *Princess Louise* was on a cruise with the submarine C27 in tow off Fair Isle. This time it was the unsuspecting U-23 which intercepted the trawler, and everything went according to plan until after the C27 took up position. Its first torpedo went wide, and the U-boat immediately broke off the attack, tried to dive

19 NA, ADM 137/201, Auxiliary Patrol Granton, w/e 4 July 1915.

20 ADM 137/201, Auxiliary Patrol Granton, w/e 4 July 1915.

21 Hurd, *History of the Great War*, 2:54–56.

but was struck by a second torpedo abaft the conning tower and blew up. Only ten German officers and crew survived.[22]

Two other U-boats managed to escape this lethal trap which also caused casualties on the British side. In August, the submarine C33 disappeared, probably succumbing to a mine, while working in conjunction with the trawler *Malta* off East Anglia, and the C29 suffered a similar fate on 29 August when it struck a mine in the Outer Dowsing field while under tow from the trawler *Ariadne*. The *Ariadne* had been disguised, fitted with a sham trawl, and involved in various special operations as far afield as the Dogger Bank for several months. On this occasion, the armed trawler and submarine had been out of sight of land for some days and, due to bad weather, the commander of the trawler had been compelled to rely on dead reckoning, but was several miles out of position. Towing a vessel anywhere near a minefield was fraught with hazard and if the submarine yawed while being pulled along then hawser between the two vessels was liable to act as a sweep wire and guide any mine encountered onto the craft.[23] Orders were subsequently issued that no submarine towing by trawlers was to be carried out within 15 miles of a minefield and in any case the practice was soon abandoned, the secret having being revealed to the enemy after the survivors of the U-23 were inadvertently allowed to mix with civilian internees about to be repatriated to Germany.[24]

A different decoy take that was sometimes adopted around this time also brought about direct action between Auxiliary Patrol trawlers and U-boats, not least on 5 June 1915 when the *Oceanic II*, one of five trawlers under the command of Lieutenant Ferguson, acted as a U-boat lure. The *Oceanic II*'s skipper, John Cowie, set a trawl as if fishing alone while the other armed trawlers maintained a distant and discreet watch. This day the bait was taken and the waiting trawlers, alerted by the sound of gunfire, dashed to the assistance of *Oceanic II*, by then exchanging shots with the U-boat. Soon Ferguson in the trawler *Hawk* came in range, opened fire, and hit the U-boat. The other armed trawlers were closing in as Ferguson dealt the lethal blow, ramming what turned out to be the U-14. All the German crew except for the commander were rescued as the vessel sank, and afterwards Ferguson and Skipper Cowie of the *Oceanic II* were awarded the Distinguished Service Cross, while a number of other fishermen received the Distinguished Service Medal. All crews on the unit's trawlers also shared a substantial financial reward from the Admiralty.[25]

Enemy submarines had taken to holding up neutral steamers in the North Sea. Therefore, Admiral Startin, SNO Granton, decided to disguise a couple of his larger trawlers, *Quickly* and *Gunner*, as small trading vessels, replete with

22 Gibson and Prendergast, *German Submarine War*, 48–49.

23 NA, ADM 137/1110, Humber, 30 August 1915 and 1 September 1915.

24 Gibson and Prendergast, *German Submarine War*, 48–49.

25 Chatterton, *Auxiliary Patrol*, 108.

false timber deck cargoes and Norwegian colours on flags. To add to the ruse, canvas covers were also carried that could be hung over the side. Startin went to sea on one of the vessels and on 20 July, an enemy submarine, evidently fooled by the deception, closed in and commenced firing on the *Quickly*, which swiftly lowered its Norwegian colours and opened fire with both a 6- and a 12-pounder gun. The U-boat was hit but managed to make off and this action reinforced the belief that there was further mileage to be had in using decoy ships to take the attack to the enemy.[26] Many so-called Q ships were to be fitted out during the war and their crews sometimes included men from the Trawler Reserve.

Subterfuge aside, there were also a few straightforward duels between armed trawlers and U-boats during 1915. In July, for example, another trawler from the Peterhead base, the *Cameo*, spotted a U-boat while on patrol in the North Sea and opened fire with its 3-pounder gun. The U-boat soon replied with its two guns, which included a much more formidable 3.4-inch armament, and tried to get around the stern of the *Cameo*. Despite clever manoeuvring by its Hull skipper, Albert (Teddy) Sayers, the armed trawler was hit three times. Shortly after engagement began, a shot from the submarine struck the fore part of the *Cameo*'s wheelhouse, passing through the steel plate, smashing the upper part of the wheel and then leaving by way of the after side open door. Skipper Sayers had a narrow escape and was wounded in his right side and eye but continued steering the ship with the remnants of the wheel. The *Cameo* was struck by a second shot which pierced the side plating on the starboard quarter just above the upper decks, then a third shot penetrated and completely wrecked the galley. The action continued for about 15 minutes with the submarine firing between 20 and 25 shots to the *Cameo*'s 11. Sayer reckoned the submarine was hit several times and certainly its forward gun ceased fire shortly after the action started. The splash of the shells was spotted by two other patrol trawlers, the *Eastward Ho* and *Sea Ranger*, and as they bore down on the scene of the action the U-boat made off at some speed. They chased the German for two hours, but the submarine used its superior speed to pull away and eventually disappeared over the horizon.[27] Sayers was also awarded the DSC and returned to the maritime front line but was tragically killed when the trawler *Gambri* was blown up by a mine in the Channel in January 1918.[28]

One of the longest Great War duels between a submarine and armed trawler also took place in July 1915 and began at 4.15 a.m. when the trawler *Pearl*, armed only with a 3-pounder gun, sighted the U-41 in thick weather off Barra Head in the Hebrides. The *Pearl* closed and tried ramming the German as both vessels exchanged fire. The U-boat was hit and attempted a crash dive while the *Pearl* tried another ramming attempt and afterwards shot away its opponent's

26 Edward Keble Chatterton, *Q Ships and Their Story* (London, 1922), 10–11.

27 NA, ADM 137/159, Auxiliary Patrol Peterhead, 20 July 1915.

28 *Hull Daily Mail*, 22 January 1918 and Commonwealth War Graves Commission.

periscope. By now the U-41 was submerged but was tracked as the trawler maintained a parallel course. The German tried shifting tack, but the *Pearl* stuck doggedly to the pursuit, even though rain reduced visibility. At 9.15 a.m. the U-boat was spotted again, quite close to the surface and the chase continued. At 11 a.m. the *Pearl* fired its explosive sweep some 500 yards ahead of the by now oily wave which marked the U-boats track. Around 1 p.m., the *Pearl* had finally to break off pursuit because of mechanical difficulties, but the U-41's problems were not at an end: that evening it was located yet again and this time pursued by the Armed Yacht *Vanessa* and the following morning two other armed trawlers also harried the craft for a couple of hours. Eventually, finally, the U-boat escaped but had to abandon its cruise and return to base sporting a badly damaged conning tower.[29]

Deployment of Anti-Submarine Nets

Anti-submarine nets were another innovation rapidly taken up in the early stages of the war that soon required the services of large numbers of fishing vessels and fishermen. Prior to the war quite substantive anti-submarine nets had been designed for deployment in the Cromarty Firth. These nets were initially made in 600 feet sections and had twelve-foot square mesh made from ⅝ inch flexible steel wire, connected by two jackstays to a number of timber floats, and were rigged to hang vertically in the sea. Each section was connected by hawser to the winch or drum of a moored trawler or drifter and some sections were rigged to be swung open as a gate to allow passage through by permitted ships.[30] The entrances to Scapa were rigged in a similar fashion by early 1915, and the idea was then extended to other principal ports and estuaries.[31]

These nets were designed for use in the sheltered channels and approaches to ports and anchorages, but the course of the war also led to the development and extensive use of offshore versions intended to try and prevent the passage of U-boats through stretches of more open water. Experiments with so-called indicator nets had taken place at Harwich and Lowestoft as early as October 1914. The basic concept of indicator nets was quite straightforward. The traditional drift nets used for fishing were replaced by 100-yard-long sections of nets constructed of high tensile wire that were initially either 30 or 60 feet in depth. These nets were often towed along and fastened together by easily detachable clips and fitted to a series of buoys on the surface. The assumption was that a passing U-boat would catch and tear loose a net section, possibly foul its propellers, and at the

29 Gibson and Prendergast, *German Submarine War*, 49–51.

30 Chatterton, *Auxiliary Patrol*, 39–41 and Marder, *From Dreadnought to Scapa Flow*, 2:73.

31 Chatterton, *Auxiliary Patrol*, 41.

same time indicate its presence and position to surface vessels by means of the movement of the buoys attached to the top of the loose net.[32]

The catalyst for the rapid deployment of the indicator net concept was the torpedoing of the pre-dreadnought battleship *Formidable*, off Start Point in South Devon on 1 January 1915 by U-24, with the loss of more than 500 crew. Clearly, a means of denying U-boats access to the Channel and beyond by way of the Straits of Dover was much needed and although the creation of substantial defensive minefields in the area might have curtailed the movement of U-boats British mines were still unreliable and apt to drag from their moorings in strong tides or bad weather. Another means had to be found and the day after the sinking of the *Formidable* the First Lord of the Admiralty, Winston Churchill, ordered four drifters: *Young Fisherman*, *Sedulous*, *Nine Sisters*, and *Ocean Comrade* to Dover. The bold idea was for the Dover Patrol to block the sea passage between England and France to U-boats by deploying curtains of nets across the Straits. Drifters, designed to lie to their nets, seemed the ideal vessels to use, as were their fishermen crews who were very experienced in the handling of their somewhat similar gear.[33]

The first trial of the indicator net drifters from Dover was carried out under the superintendence of Rear Admiral, the Hon. Horace Hood, and began on 15 January 1915. It was not long before these four fishing vessels were joined by many others. By June 1915, the Dover net drifters had reached a maximum strength of 132 drifters and three yachts, manned by upwards of 1,500 officers and men.[34] Eventually large numbers of drifters were also deployed on indicator net duties in waters around the United Kingdom and much further afield. Dover led the way in the deployment of this method of anti-submarine warfare and large numbers of drifters were soon arriving at the port, many initially from Yarmouth and Lowestoft, while others were drawn from further afield, particularly the fishing ports on the North-east coast of Scotland. Substantial Admiralty orders for steel wire nets were placed with manufacturers (Figure 6).

Large numbers of herring fishermen, many just recruited into naval service, descended on Dover with their drifters. Strangers to naval discipline, their outlook, attitudes, and often well-worn fishing apparel and ships, initially proved something of a surprise to those Dover Patrol officers and men stationed at the port who were accustomed only to Senior Service standards of smartness and discipline, but the work of trialling, laying, and working the nets continued as both sides came to terms with each other.

But against the backdrop of an impending initial German declaration of unrestricted submarine warfare, there was little time to stand on ceremony. Naval officers and fishermen alike had to work closely together to overcome the

32 Hurd, *History of the Great War*, 1:369.

33 Chatterton, *Auxiliary Patrol*, 39–41 and Corbett, *Naval Operations*, 2:271.

34 R. Bacon, *The Dover Patrol*, vol. 2 (London, 1919), 416.

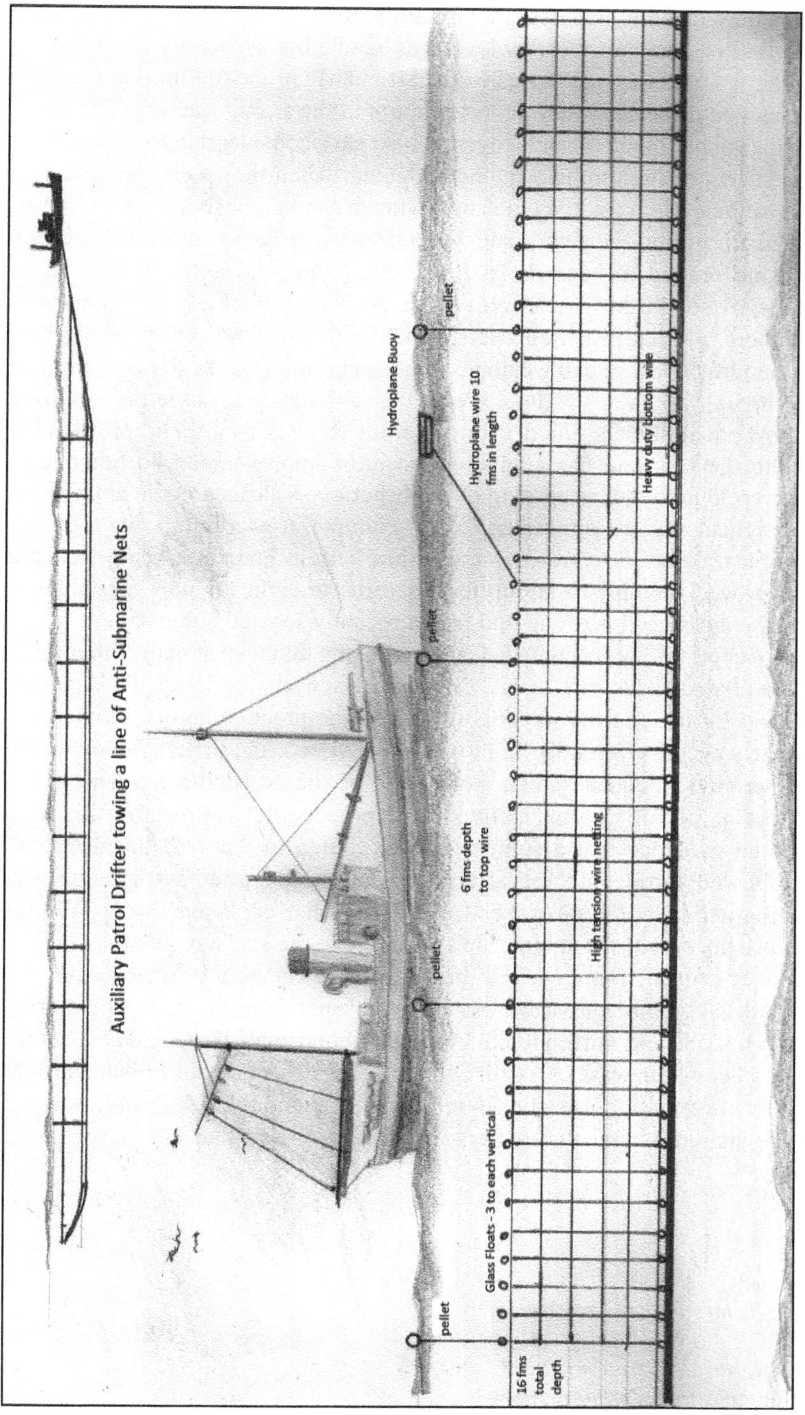

Figure 6: Auxiliary Patrol drifter and anti-submarine nets, *c.*1918
Source: Alan Hopper

difficulties they faced in deploying and maintaining these curtains of steel wire on the high seas. The nets were awkward to handle but especially cumbersome to lay in the sea, and the steel wires were particularly prone to kinking when the current changed direction every six or so hours in the strong tideway. The quick release clips which linked the nets together also gave considerable trouble as they had to be strong enough to hold the nets together when they were being hauled in and yet sufficiently weak to come apart when a submarine met the nets. It also proved difficult to find the right material to keep these heavy nets afloat. Kapok was tried but tended to become sodden all too soon. Glass balls were more buoyant and although they sometimes broke away they were eventually used in large numbers.[35] Much time and energy could also be expended in laying nets only for them to be lost in bad weather; on one occasion at least 90 nets were lost during a three days' gale.[36] Although working with nets was something entirely new to most naval officers, the driftermen were able to bring their considerable expertise to the task and Hurd tells us that most improvements in practice in this field were due to the harnessing of the fishermen's skills and the application of their practical suggestions. Initially, for example, it was found that a drifter could only lay about 300 yards of nets within half an hour in a heavy sea, but in time it proved possible to lay up to 800 yards in eight minutes, particularly useful if attempting to surround and trap a recently located submarine.[37]

The concept was soon adopted elsewhere, being taken up at many other ports around the British Isles, not least Peterhead, Harwich, Portland, Portsmouth, Milford, and Larne, and the impetus to diffuse the practice gained momentum from an early success. Around 12.30 p.m. on 4 March, the drifter *Robur*, while lying to her nets about six miles north-east of the North East Varne Buoy, noticed that a line of five net pellets were proceeding against the tide at 4 knots, strong evidence that a submarine was fouled in them. A hot destroyer pursuit followed throughout the afternoon and shortly after a periscope was spotted around 5 p.m. the destroyer *Gurkha* fired her explosive sweep. Within 30 seconds a portion of the submarine's stern broke the surface then the *Gurkha* and fellow destroyer *Mauri* shelled the conning tower as it emerged from the depths. With all chance of escape gone, several members of the German crew came out on deck and surrendered. All the submariners were rescued before U-8 slipped below the waves.[38] Although the net drifters had not finished off the U-boat their contribution was considered invaluable and the Admiralty awarded £500 to be shared among the drifters and trawler crews that had participated in the hunt.[39]

---

35 Ibid., 418.

36 Hurd, *History of the Great War*, 1:383.

37 Ibid.

38 Corbett, *Naval Operations*, 2:276.

39 Chatterton, *Auxiliary Patrol*, 66–67.

The sinking of U-8 seemed to justify the use of the indicator nets and considerable resources in terms of fishing vessels, ships, backup, and supplies were poured into the project. Long lines of drifters and nets were deployed to try to close off the Straits of Dover and various harbours and bays where it was thought submarines might seek to shelter and recharge were also netted. Hundreds of drifters were deployed in waters off the westward coasts of the British Isles, many to try and create barriers across the St George's Channel and the North Channel, the two entrances to the Irish Sea.[40] Those sent to Larne to work the North Channel between Ireland and Scotland towed nets backwards and forwards across an area of sea some 30 miles long by 22 miles wide, while armed trawlers patrolled in the waters at either end, the idea being to grant no respite to any exhausted U-boats that tried to surface after a gruelling voyage under the netted area.[41]

An early German response to the widespread deployment of indicator nets was to fit net cutting gear to the bows of their U-boats. Though further U-boats did fall foul, and nets occasionally helped in their destruction, the return for the massive level of effort put into the original indicator net strategy around the British Isles seems in retrospect to have been scarcely justified by the outcome. In terms of providing a barrier, they remained difficult both to deploy and keep on station. To a degree, their deployment created an initial false sense of security, for it was later apparent that on many occasions they had offered no substantial impediment to the passage of U-boats, although later innovations such as the fitting of mines to the nets and their deployment, particularly when towed, and used in combination with hydrophones and depth charges in submarine hunts, meant they continued to be utilised in combating enemy undersea depredations throughout the rest of the conflict.

Dover Patrol operations to throttle the flow of U-boat traffic from the Belgian coast extended far beyond these nets in the first three years of the war. In early September 1915, Admiral Bacon used monitors and many auxiliary craft to bombard the low-lying Belgian coast, and his flotilla of net drifters distinguished themselves while under return fire from shore-based batteries. In April 1916, Admiral Bacon began laying a Belgian coast barrier with the idea of restricting passages for U-boats. Once laid, drifters, supported by destroyers, were expected to keep it in repair. The barrage was complete by early May and maintained until the following October. Though patrolled during the day, it was left unattended after dusk and the Germans usually found little difficulty in getting through this at night. Though it is generally thought to have been largely ineffective, associated operations did lead to the destruction of a UB-3 during that first April. The drifter *Gleaner of Sea* was lying to nets off the Thornton Ridge when a U-boat fouled the wires and had to surface. The drifter's skipper, Robert Hurren,

[40] Hurd, *History of the Great War*, 1:380–381.

[41] Ibid., 382.

pitched first one lance bomb at the German, which caused serious damage, then followed with a second. Another drifter, EES, came up and dropped more bombs and marked the spot with a dan buoy. Skipper Hurren also summoned the destroyer *Alfredi*, which fired its sweep over the spot.[42] The U-boat had been destroyed and Skipper Hurren and his mate were awarded respectively the DSC and DSM while the crew shared £1,000.

This coastal net barrage was left unattended throughout the winter but was reconstituted a little further offshore the following year as part of a projected allied landing on the Belgian coast which came to nought. The logistical problems related to trying to maintain this would-be barrier meant that of itself it continued to cause only limited irritation to German U-boat traffic in the area.

Hostile Actions and Motor Launch Reinforcements

By the end of 1915 the growing Auxiliary Patrol fleet of hired armed trawlers and drifters was being reinforced by the arrival of fast wooden motor launches. A total of 550 of these vessels were ordered, initially in July 1915 from North America, and the first deliveries were made in October and were soon in commission. Most of these craft were 80 feet in length and fitted with a 13-pounder gun and had the advantage over trawlers of high speed and shallow draft. Although they lacked the reliability and seakeeping capacities of the fishing vessels they were nevertheless eventually seen as a welcome addition to the Admiralty's armoury in this increasingly bitter war. There was, however, a further cost: it proved necessary to train up an additional 2,500 skilled mechanics to operate and maintain their engines.[43] Yet the motor launches were soon in service and by 1916 most Auxiliary Patrol bases around the British Isles had at least a flotilla of these craft.

Attacks on fishing and merchant vessels continued in 1916, and despite the arrival of the motor launches the resources of the Auxiliary Patrol were continually stretched. Apart from carrying out regular anti-submarine patrols along seaways and in bays, there were increasing calls for trawlers to act as escort vessels and to respond to some serious submarine attacks on fishing vessels further out to sea. Gradually, the overall firepower of the anti-submarine patrol trawler fleets improved. More trawlers were fitted with 6- or 12-pounder guns, while many of their original 3-pounders were fitted to previously unarmed drifters. But even the largest of these guns did not match the firepower of the submarines on the surface, and with innovations like the depth charge and hydrophone still in the early stages of either development or deployment the armament available to most of Auxiliary Patrol vessels when trying to pursue submerged U-boats throughout

42 Bacon, *Dover Patrol*, 2:430.

43 Chatterton, *Auxiliary Patrol*, 104.

much of 1916 remained largely limited (as outlined in Chapter 2) to the explosive sweep and the lance bomb.

There were many hostile actions, even pitched battles, fought between armed trawlers of the Auxiliary Patrol and enemy submarines during the year, and the waters off the Scottish coasts were the scene of many such encounters. In May 1916, Rear Admiral Simpson, SNO Peterhead, ordered his armed trawlers to pay attention to sections of the Tyne to Bergen trade route. Simpson anticipated further enemy activity in the area, even though the Germans had ostensibly suspended their attacks on commerce following pressure from the USA. In fact, German submarines had been ordered to sweep through the northern regions of the North Sea and take up positions off the Humber, Firth of Forth, and Scapa Flow in the hope of catching British naval forces at sea. The minelaying U-boat, U-74 was apparently part of this plan, and was following orders to lay its cargo off the south-east corner of the Firth of Forth in the hope of ensnaring part of Admiral Beatty's Rosyth-based battle-cruiser fleet. However, on 27 May 1916, the submarine was spotted some 100 miles to the seaward of Peterhead by the armed trawler *Searanger*, commanded by Lieutenant H. J. Bray, RNR, which was at sea with three other Auxiliary Patrol vessels, the *Oku, Rodino*, and *Kimberley*. Both sides had clearly engaged in some sort of subterfuge: U-74 had set a sail and was making smoke to try and disguise its true self in order to get close to potential targets while the four armed trawlers were cruising in no obvious formation, seeking to resemble a scattered group of unarmed fishing vessels in order to try and lure a U-boat into attacking. In this case, it was U-74 that seems to have been taken in and its commander did not discover the ruse until hit by a shell from the *Searanger*: then the trawlers *Oku* and *Rodino* joined in the attack. The U-boat was well armed and opened fire with its 4.1-inch and 22-pounder guns but sustained more damage from the trawlers as they closed and its after gun was put out of action. Eventually, the vessel ceased firing altogether and tried to dive, but was by then half encircled by the trawlers as shot after shot hit home. The damage they wrought must have been substantial. Instead of submerging, the German vessel rose high in the water and then, listing heavily to port, it headed erratically in the direction of the *Kimberley*. The *Oku* and *Searanger* both attempted to ram the stricken vessel which finally disappeared below the surface following a last shot from the *Kimberley*, leaving only a coating of oil to mark its final descent.[44]

This action preceded the clash of great warships at the Battle of Jutland by just four days, and although Auxiliary Patrol vessels had no part in this distant fleet action they played a role in repelling enemy submarine activity as vessels from the Grand Fleet, some badly damaged, made their way back to British bases. Most notably, the armed yacht *Mingaray* and its Admiralty trawler unit

---

44 Gibson and Prendergast, *German Submarine War*, 98–99 and Chatterton, *Auxiliary Patrol*, 150–151.

from Granton did valuable work in chasing off a submarine apparently waiting to intercept the much-damaged HMS *Warspite* as it returned to the Firth of Forth.

Meanwhile, back on 21 April 1916, a few days before the Easter Rising in Dublin, the armed trawler *Lord Heneage*, on patrol off the south-west coast of Ireland, chased a suspicious, ostensibly Norwegian steamer, the *Aud*, which made off as it approached. The *Aud* moved off at full speed, making 13 knots, and managed to stay well ahead of its pursuer, but the armed trawler, under Commander William Bee, maintained contact and successfully shadowed the ship all afternoon and radioed its position to the Admiralty. Around 6 p.m. two Royal Navy sloops joined the hunt and after a flotilla closed down on the steamer its captain was obliged to hove to. The steamer turned out to be carrying a huge consignment of guns brought from Germany intended to be distributed to volunteers across Ireland after the planned insurrection, but they didn't reach land. As its German captain Kurt Spindler was escorted by the warships into Cork harbour he ran up two German ensigns and blew the bottom out of his ship, which sank straightaway. Spindler may have been making an attempt to block the approaches to Cork harbour with this action, but he did not succeed. More importantly for the future history of Ireland, the crucial consignment of arms sent by the Germans now lay on the seabed and out of reach. When plans for the Easter Rising had initially been devised, the idea had been to seize and hold points across Dublin until the arms brought by the *Aud* could be distributed across Ireland. The Loss of the *Aud* meant there were no substantial supplies of weapons available for armed uprisings elsewhere, but despite the odds being stacked against them the rebels went on with the uprising. They were now doomed to an early military defeat, but the very hopelessness of their action probably proved a long-term benefit to their cause. Their actions and the subsequent heavy-handed retaliation by the British inflamed the Irish political situation and ignited outright nationalist feelings, the Easter Rebellion proving such a major watershed in Irish History.

There was a further dimension: both the steamer and the armed trawler that pursued it were from the same peacetime port. The *Aud* was originally the *Castro*, a Wilson Line steamer built and operated out of Hull. It had been interned by the Germans at the outbreak of the war, initially renamed the *Libau*, and then disguised by the Germans as the Norwegian steamer *Aud* for the gun-running voyage. The *Lord Heneage* was in peacetime a Hull trawler that had been requisitioned and armed by the Admiralty after the outbreak of war.[45]

During the summer of 1916, German U-boats focused a great deal of attention on trying to reduce Britain's North Sea fishing capability. On many occasions, substantial numbers of unarmed fishing craft were sent to the bottom. Twenty-five fishing vessels were sunk off the Tyne during July, while other vessels were lost off the Yorkshire coast, the Dogger Bank, Shetland, and

45 Robinson, *Far Horizons*, 71–74.

other places; a further heavy attack on 24 September by Carl-Siegfried Ritter, Commander of U-57, and detailed in Chapter 6, resulted in the sinking of 13 Scarborough trawlers. Their crews survived but the port's existing fishing fleet was virtually wiped out.[46]

Meanwhile, the armed trawlers of the Peterhead Auxiliary Patrol units remained very much in the front line, taking part in a series of actions in the North Sea and beyond. On the morning of 7 July, a Peterhead armed trawler group caught sight of first one and then a second U-boat in the vicinity of 58.20 North, 0.48 East. Two trawlers, the *Glamis* and *Prince Consort*, pursued the first enemy craft for 20 miles before finally losing contact. The second U-boat dived after being attacked by the armed trawlers *Martin*, *Albatross*, and *Editor*. The armed trawlers regrouped and at 7 p.m. on the same day made a further U-boat sighting just six miles from the morning's action. This time, the *Albatross* got close enough to open fire at around 8 p.m., but the U-boat replied, closing with the trawler and opening fire with her superior armament. The submarine inflicted considerable damage on the Auxiliary Patrol vessel before the armed trawler *Martin* arrived on the scene and joined in the action. The U-boat then turned both guns on the *Martin* and the exchanges continued until a shot from the *Albatross* appeared to hit the German vessel, which began making smoke and made off at speed eastwards while continuing to fire towards its opponents. The trawlers followed but lost contact around 9.40 p.m. Five hours later the group made yet another sighting, this time a U-boat with its wireless masts up, but this too sped off and submerged after being shelled.[47]

The Auxiliary Patrol units were still poorly armed when up against U-boats, and while more now had 12-pounders, some still only deployed nothing more powerful than 3-pounder guns. Later events suggest that the 7 July actions were part of a German plan to make a concentrated attack and wipe out their armed trawler adversaries. Certainly, another unit of armed trawlers from Peterhead, the *Era*, *Nellie Nutten*, and *Onward* were engaged by U-boats in another extended North Sea action, just a few days later, this time some 120 miles south-east of Girdleness.[48] Though based at Peterhead while on war service, the vessels involved in this latter attack were registered at Hull and Granton.[49]

Events opened when Lieutenant Claude Asquith, RNR, leader of the unit, on HMAT *Onward* signalled the other two trawlers that he had sighted a submarine. The *Nellie Nutten*'s skipper, 50-year-old Charles Angus, in peacetime a fisherman from Newhaven near Edinburgh, immediately headed for the *Onward* which

---

46 Robinson and Hart, *Viola*, 99–100.

47 Hurd, *History of the Great War*, 2:277–278, 279.

48 Henry Newbolt, *History of the Great War*, vol. 4, *Naval Operations* (London, 1928), 348.

49 Gill, *Lost Trawlers of Hull*, 83 and 'World War 1 at Sea – Ships of the Royal Navy, 1914–1919, Fishery Port Registrations, Part 2, by vessel name': www.naval-history.net/WW1NavyBritishShips-Dittmar6FishPortNos2.htm (accessed 3 October 2017).

had by now engaged the U-boat and opened fire. Whilst the *Onward* had a 12-pounder gun, the *Nellie Nutten* possessed nothing more powerful than a 3-pounder and was completely outranged. Two more U-boats then appeared and joined in the fray and soon afterwards the three German craft, apparently the U-24, U-52, and U-69, positioned themselves to the port and starboard of the Auxiliary Patrol unit and inflicted a long-distance and often accurate pounding on their opponents. The trawlers, although completely outranged, continued to return fire but to no avail. It proved a once-sided affair: a couple of hits, one of which passed through the stern, disabled the *Nellie Nutten.* The crew sought help but could expect nothing from the *Era* which they could still see in the distance, but had by now had been closed by two of the U-boats and was on fire from end to end, The *Nellie Nutten*'s crew managed to steam their stricken vessel towards a Dutch fishing lugger, the *Doggerbank* of Scheveningen, and got aboard before their trawler went down. In the distance, they caught sight of the *Onward*, by now also on fire and seemingly out of control. Eleven of the 12 members of the *Nellie Nutten*'s crew survived their ordeal and were landed by the Dutch in Aberdeen, while the crews of the *Onward* and *Era* were taken prisoner. Those captured included Commander Asquith of the *Onward*, who was awarded the DSC for gallantry.[50]

The 59-year-old Hull skipper of the *Era*, George Noel, was also taken prisoner but died of his wounds, possibly a severed hand, two days after the action, in hospital in Emden,[51] and was buried in Hamburg. Skipper Angus of the *Nellie Nutten* was also awarded the DSC for the 'skilful and seamanlike manner in which he manoeuvred his vessel when she was disabled by gunfire, thereby saving practically the whole of his ship's company'.[52] In October 1918, another member of the *Era*'s crew, Frederick James Tenley, was also awarded the DSM for his actions that day. The attacks certainly illustrated that 3-pounder guns were of no use against well-armed and well-operated U-boats, and spurred on a programme of increasing the armament deployed on Auxiliary Patrol vessels.[53]

The attacks on Auxiliary Patrol vessels continued. On 27 September, the armed yacht *Conqueror* II and Auxiliary Patrol Trawler *Sarah Alice*, in peacetime a Fleetwood vessel, were torpedoed while investigating what turned out to be a British steamer. Three torpedoes were evidently fired, the first of which struck the *Sarah Alice*, and although the second passed under the stern of the *Conqueror* II, a third found its target in the magazine before the fore-bridge. There were no survivors from the *Sarah Alice* but 17 officers and men from the armed yacht were picked up from Carley Floats by the destroyer *Sylvia* the next day. The survivors later recalled how the submarines remained on the surface

---

50 Hurd, *History of the Great War*, 2:278–279.

51 *Hull Daily Mail*, 5 September 1916.

52 *Liverpool Daily Post*, 26 October 1916.

53 Newbolt, *Naval Operations*, 4:348.

near the floats after dark and fired two Very Lights, presumably to attract some vessel to the spot.[54] A seemingly compassionate act in what was by now a vicious war. The sinking of the *Sarah Alice* and *Conqueror* II was attributed to Commander Hans Walther of the U-52, who was also credited with having sunk the *Onward* in the action in July off the west coast of Scotland.[55]

Another Orkney-based armed trawler sunk in action with a German U-boat around the same time was the trawler *Orsino*, which in peacetime was a Hull North Sea boxing fleet trawler operated by the Hellyer Steam Trawling Company. The *Orsino*, sister ship to *Viola*, which survives to this day, had been requisitioned in September 1914 and was a member of one of the Auxiliary Patrol units, working from Orkney. The vessel was last seen by the trawler *Sesostris* as it headed out to sea from Loch Eribol on its way to Stromness on 29 September at 5.30 a.m. but never made land again.[56] Nothing more was heard about the vessel for some time, although the portion of the ship's boat, embedded with shell fragments, that was subsequently picked up in Ireland Bay, near Stromness, indicated the vessel had met a violent end. This was finally confirmed at the end of October when the wives of two crew members received post from their spouses, by then held in a German prison camp at Brandenburg, to say that they along with two others were now prisoners of war.[57] Their vessel had encountered the U-55, and after a fierce fight the trawler had been sunk north of Strathy Point on the Pentland Firth. All others on board had been lost along with the vessel but the survivors were picked up by the submarine and taken back to Germany.

Given the nature of wartime geography and the strategic importance of the waters they served in, all the vessels which comprised the Dover Patrol, particularly the armed trawlers and drifters, were all too often subject to enemy attack. The initial raid on Dover Patrol forces by German destroyers from the occupied Belgium ports took place during the night of 26–27 October 1916 and led to the sinking of six drifters, three others being badly damaged. Forty-five crew were said to have been lost during the raid and another ten taken prisoner. A month later, on the night of 23 November, the vessels of the Dover Patrol were subject to a second though largely unsuccessful destroyer assault and faced further such raids during 1917. These were followed by aircraft attacks and the area's forces remained very much on the front line almost to the very last days of the war.[58]

---

[54] NA, ADM 137/435, Auxiliary Patrol Orkney and Shetland, 20 September 1916.

[55] In between these dates he is also attributed with the sinking of the light cruiser *Nottingham*; see uboat.net: https://uboat.net/wwi/ships_hit/4475.html (accessed 3 April 2017).

[56] NA, ADM 137/435, Auxiliary Patrol Orkney and Shetland, 1 October 1916.

[57] NA, ADM 168/232, 2 January 1917.

[58] Chatterton, *Auxiliary Patrol*, 165–168 and 187–192.

Admiralty-Built Trawlers and Drifters, Hydrophones, and Depth Charges

It was proving increasingly difficult for the Admiralty to requisition more trawlers or drifters without seriously compromising the country's remaining fish supplies and nationwide food shortages had become a serious concern. Indeed, by September 1916, about 1,300 trawlers had been requisitioned leaving only around 500 in private hands.[59] The Admiralty's response was to order in October 1916 the construction of another 250 trawlers built to a set of three standard designs by the leading builders: these were known as the Mersey, Castle, and Strath classes (respectively 138½, 125, and 115 feet in length) while they also took up many fishing vessels currently on the stocks and intended for private trawling companies. The first of these standard classes of trawlers came into service in early 1917. Orders for an additional 250 Admiralty trawlers were also approved by the War Cabinet in 1917. Additionally, the Admiralty placed orders for over 150 drifters. Indeed, by the end of the war, orders for the building of 550 trawlers and 335 drifters had been approved, although not all were in the end constructed.[60] Certainly, the new vessels were a welcome reinforcement for the hard-pressed Auxiliary Patrol anti-submarine and minesweeping flotillas. Many were refitted for fishing when sold on after the return of peace.

But what of the weaponry? Gunnery shortcomings in surface engagements with U-boats were certainly a feature in the loss of the three Peterhead Auxiliary Patrol trawlers in July 1916, and, at Admiral Jellicoe's own suggestion, an additional 57 12-pounder guns were distributed and fitted to east-coast armed trawlers in the following months.[61] Many more were made available for Auxiliary Patrol vessels during the last two years of the war. Further technological developments, some of crucial value in the war against U-boats, were to come increasingly on stream from the closing months of 1916 and were to enhance the effectiveness of the Auxiliary Patrol vessels when pursuing their underwater prey. One major development was the hydrophone.

Commander C. P. Ryan, RN, pioneered British work on the hydrophone. A wireless specialist, he had left the Navy but returned in 1914 and began working on submarine detection. After attracting official support for his research in 1915, he established a base at Hawkscraig, Aberdour in Fife and was allocated the drifter *Tarlair* for sea trials of his device. Initially, a coastal chain of listening devices was established, and its success led to the development of hydrophones that could be used on ships. During the remainder of the war various improvements were made and alternative and more effective designs were developed.[62] The first

[59] Friedman, *Great War at* Sea, 327.

[60] Ibid., 327–328.

[61] Hurd, *History of the Great War*, 2:279.

[62] Friedman, *Great War at Sea*, 292.

hydrophones could only indicate that a U-boat was somewhere in the vicinity, but in 1917 new versions and methods of deployment made it possible to indicate the direction of the U-boat with somewhat greater accuracy.[63] During the autumn and winter of 1916, no fewer than 1,500 hydrophone drifter sets were supplied to the Auxiliary Patrol.[64] They were deployed by the service in ever-increasing numbers during the following year and by the middle of March 1917 it was reported that all Auxiliary Patrol vessels operating in Patrol Area IX Humber had already been fitted with a hydrophone.[65] Many of the new Admiralty class trawlers came into service already fitted out with hydrophones.

Initially, those who were earmarked to operate the devices off the British coasts attended training courses at Hawkscraig. By the end of the war, 1,000 officers and 2,371 ratings had either attended Tarlair training courses or received instruction courses from Tarlair's peripatetic training parties. Many other sailors were probably instructed in a cascade method as in the Humber area. Shortly after hydrophones had been fitted, the skippers of every Auxiliary Patrol vessel were ordered to select two men to become hydrophone listeners and these were to be trained by Mr Pratt, a gunner, who had recently attended the practical listening course at the Fife base.[66]

Aberdour, or Tarlair, eventually deployed 12 requisitioned drifters which were used for cable-laying to fixed listening stations.[67] Hydrophones had numerous shortcomings, not least the problem of needing to stop engines to listen, which made the vessel potentially more vulnerable to attack. U-boats were also soon fitted with hydrophones as well and they too listened for the sound of the vessels hunting them on the surface. On occasions, deadly games of cat and mouse ensued: vessels on the surface and U-boats below, each trying to outmanoeuvre the other. Hydrophone hunting vessels also had to contend with the engine noises from the other craft engaged in hunting U-boats.

Another new weapon, which proved to be extremely important in the Auxiliary Patrol's war against U-boats, was the depth charge. The first of these devices, which came to consist of powerful explosive charges packed in steel drums and detonated by a pressure detonator that could be set to explode at a given depth, appeared in 1915, but the early versions were of only limited effectiveness and available in small numbers. It has been claimed the disguised armed trawlers *Gunner* and *Quickly* were the first anti-submarine vessels to deploy depth charges in anger in their attack on a U-boat in July 1915.[68] Originally, depth charges were merely dropped over the side, but in August 1917 a throwing mechanism that

---

63 G. C. Peden, *Arms, Economics and British Strategy: From Dreadnoughts to Hydrogen Bombs* (Cambridge, 2007), 57.

64 H. W. Wilson, *Hush or the Hydrophone Service* (London, 1920), 7.

65 NA, ADM 137/943, Auxiliary Patrol Humber, 17 March 1917.

66 Ibid., 31 March 1917.

67 Wilson, *Hush or the Hydrophone Service*, 10.

68 R. Compton-Hall, *Submarines at War, 1914–1918* (London, 1970), 101 and

could pitch the depth charge some 40 or so feet away from the attacking craft was introduced. Before June 1917, anti-submarine vessels carried no more than two depth charges, but from August the allocation was increased to six. Monthly production was steadily stepped up: output in July did not exceed 600. Two thousand were turned out in October, and this figure more than doubled again by the end of the following half-year. The throwing mechanism also made it possible to deploy heavier charges and by the end of 1917 improvements to the pressure detonator meant they could be used at depths of up to 200 feet.

Nonetheless, in early 1917, the depth charge was in many respects still being developed and diffused as a potential weapon of war for the Auxiliary Patrol and other branches of the Royal Navy. The numbers used in attacks by all branches of the Royal Navy in every maritime theatre of war during 1917 never exceeded 300 a month, whereas in the last six months of the war more than 2,000 a month were being used.[69] No more than three U-boats were sunk with depth charges in 1916, and just twelve in 1917, but the device accounted for more than double this latter number over the last ten months of the war.[70]

1917: Armed Fishing Vessels and the Resumption of the German Unrestricted Submarine Warfare Campaign

By late 1916, Britain's own naval blockade of enemy ports was having a serious impact on the German economy and society, causing serious shortages of food and raw materials. With the continued stalemate on the western front and the failure of either side's fleet of dreadnoughts to deal a knockout blow at the Battle of Jutland in the early summer of 1916, there was a strong feeling in some circles of the German High Command that the resumption of unrestricted submarine warfare was the only way to bring the war to an end before their country's ability to maintain the struggle was fatally weakened. A German study had calculated that if an average of 600,000 tons of shipping could be sunk every month then Britain's ability to wage war would be degraded, insufficient vessels would be built to replace those lost, and the mercantile fleet would be much diminished, leaving the country with no option but to sue for peace within six months. The debate amongst the German leadership surrounding this idea was intense because it was also recognised that such action would undoubtedly bring the USA in on the side of the allies. But those favouring the resumption believed the war could be ended before America could raise, train, and transport sufficient

H. P. Wilmot, *The Last Century of Sea Power*, vol. 1, *From Port Arthur to Chanack* (Bloomington, Ind., 2004), 1894–1922, 399.

[69] Arthur Marder, *From the Dreadnought to Scapa Flow*, vol. 4, *1917, Year of Crisis* (Oxford, 1969), 71–72.

[70] Crompton-Hart, *Submarines at War, 1914–1918*, 101.

troops to Europe to make a decisive contribution. Many civilian politicians in Germany did not believe that the U-boats could achieve the predicted outcomes, but eventually the military hawks won the debate and on 1 February 1917 the campaign of unrestricted submarine warfare in the waters around the British Isles was resumed and was to last for the final 21 months of the war.[71]

This long, last, and bitterly fought U-boat offensive proved a bleak struggle for everyone, belligerents and neutrals alike. This was no rerun of Trafalgar, it was a modern struggle, a precursor of things to come, a dirty war of attrition, gauged by performance indicators and targets. Although the resumption of the campaign brought in the Americans, as predicted – they declared war on 1 April 1917 – for the first few months of the campaign the German hawks must have thought their decision was vindicated by the outcomes. In the same month the Americans joined the war effort, 841,118 gross registered tonnage (grt) of allied and neutral shipping was sunk, and in June the figure was 669,218 grt.[72]

In part, the Germans achieved this through an intensive increase in submarine minelaying activity all around the British coasts as outlined in Chapter 3, but direct attacks by U-boats on merchant craft with torpedoes, gunfire, and the like took a greater toll. Gradually, slowly, the tide began to turn back towards the allies. The tremendous April losses sustained by allied and neutral mercantile fleets were never surpassed, and June 1917 proved to be the only other month when the Germans exceeded their 600,000 target. Nevertheless, attrition rates remained extremely high, with a monthly average 451,000 grt being sent to the bottom in the third quarter of the year, although this fell to 365,000 grt in the final three months.[73]

Several measures assisted the allies to counter this U-boat campaign. By then, the fishermen and other seamen of the Auxiliary Patrol who remained in the forefront of the action in this renewed German offensive were well trained, most were battle hardened, a substantial number having served for more than two years on the maritime front line. New technology also helped: many vessels had deployed hydrophones and valuable experience was being accumulated in terms of the best ways of making effective use of these listening devices, while Auxiliary Patrol reports also show armed trawlers and the like gradually made more use of depth charges when taking on U-boats as they became increasingly available over the year.[74]

---

71 Newbolt, *Naval Operations*, 4:352.

72 Jurgen Rhower, 'Submarine Warfare: Central Powers', in Spencer Tucker (ed.), *The European Powers in the First World War: An Encyclopaedia* (London, 1996), 671.

73 Ibid.

74 See, for example, NA, ADM 137/539, weekly reports, July–December 1917.

Fishermen, Fishing Vessels, and Convoys

However, the most effective innovation of 1917 was probably the introduction of convoys. Although historically convoys had been used in many major conflicts, the Admiralty had hitherto been opposed to the convoy in the Great War, but in December 1916, in order to encourage a number of Scandinavian shipowners who had suffered heavy losses from submarines to continue trading with Britain, a series of so-called protected sailings for their ships were instituted. From this time forwards, Auxiliary Patrol trawlers began to escort such shipping across the North Sea and along the British coast and the convoy concept was duly extended, even though this meant diverting the much in demand patrol vessels from other duties. During 1917, larger merchant ships crossing the Atlantic or voyaging to the Mediterranean began to be escorted by American as well as British destroyers and warships. Much North Sea convoy work involved Auxiliary Patrol vessels, and initially Lerwick became an assembly point or clearing house for vessels voyaging to or from the Humber or readying themselves to cross the North Sea.[75] At first, some 45 armed trawlers and around 20 destroyers were deployed in these tasks in the Orkney and Shetland area.

Convoy escort duties drew in Auxiliary Patrol vessels and other naval craft from bases up and down the east coast of England and Scotland. The strain this placed on the already stretched resources was clearly being felt in the Tyne area, and it became difficult to cover all escort duties without reducing anti-submarine patrols. The situation became so acute that virtually no patrolling was done there in the last week of March 1917. The Humber Auxiliary Patrol resources were also much stretched as the extension of the Shetland convoy support duties also coincided with plans to escort fleets of civilian trawlers voyaging to and from the Icelandic fishing grounds. These problems continued into the summer, the SNO Tyne noting in August that his armed trawlers were now seldom available for patrol work due to their employment on escort duties.[76]

Trawlers remained in demand for this escort work because of the dearth of other vessels that possessed their seakeeping and endurance capacities. The SNO Humber, for example, considered that the ageing 30-knot destroyers also allocated as convoy escorts were only really suited to such duties when the weather was calm, and he feared that some may founder if caught away from port in bad weather; they were also beset with many more mechanical problems than the much more robust trawlers. On one occasion, around this time, two 30-knot destroyers escorting one of the convoys from the Humber to Shetland were obliged to run for shelter in Aberdeen when weather conditions deteriorated, leaving the convoys to proceed northwards with escort trawlers only.[77] Whalers

---

75 Chatterton, *Auxiliary Patrol*, 217.

76 NA, ADM 137/539, Auxiliary Patrol Tyne, 13 August 1917.

77 NA, ADM 137/719, Auxiliary Patrol Humber, 19 May 1917.

were also used for escort work, but they too had serious shortcomings in terms of their limited bunker capacity, high fuel consumption levels, and maintenance requirements. Whilst the trawlers carried sufficient coal for two round trips from the Humber to Shetland, the whalers had to refuel after one voyage out and back, and were obliged to stand down for routine maintenance, boiler cleaning, and the like, every four weeks, whereas the trawlers could carry on working for up to six, even eight weeks without requiring such attention.[78]

The pressure on the patrol trawler crews was intense and the SNO Tyne strongly resisted any attempt to increase the sea time of his Auxiliary Patrol vessels and personnel to cope with demands for their services. The crews under his command were at that time working six days at sea followed by three days ashore. He considered any extension could not be achieved without seriously affecting the health of the crews.[79]

Despite the pressure of work, Auxiliary Patrol trawlers and crews based on the Humber managed to fulfil their new escort duty commitments. Marder states that the first full Scandinavian convoys left the Humber and Lerwick on 29 April,[80] and as early as mid-May it was noted that six escort forces of armed trawlers had conveyed 71 vessels northwards and 30 in the opposite direction. The diversion of vessels from anti-submarine patrols was a considerable problem for these north-east coast Auxiliary Patrol areas at this time because this was a period of transition. While vessels bound for Scandinavia were convoyed, other merchant vessels were still voyaging alone up the coast. The problem for the Auxiliary Patrol vessels in trying to guard convoys and at the same time protect unescorted vessels became somewhat easier when the convoy system was later extended to cover all vessels voyaging along the North Sea coast, although orders prohibiting unescorted sailings from north-east ports were rescinded again for a time in November.[81]

The development of safe convoy practice was accompanied by an array of logistical problems. Apart from the inevitable difficulties that arose when trying to keep a collection of vessels of different sizes and speeds in some sort of formation on the high seas for long periods, the whole issue of assembling or joining convoys presented formidable challenges. The SNO for Tyne Auxiliary Patrol area, for example, needed to ensure merchant ships sailing from the north-east ports were ready to join passing convoys when they arrived, but this proved difficult unless they were kept waiting out at sea, beyond the safe confines of harbour, and thus exposed to potential U-boat attack. The ports of Blyth, Sunderland, Hartlepool, and Tees were also to some extent tidal and this posed major problems as it was not possible to adhere to convoy arrival and departure timetables unless the tide suited.[82]

---

78 Ibid., 4 August 1917.

79 NA, ADM 137/539, Auxiliary Patrol Tyne, 13 August 1917.

80 Marder, *From Dreadnought to Scapa Flow*, 4:144.

81 NA, ADM 137/539, Auxiliary Patrol Tyne, 3 November 1917.

82 NA, ADM 137/438, Auxiliary Patrol Humber, 28 April 1917.

With the onset of winter, convoy timings were also increasingly disrupted by the weather, which created further problems. It was noted in December, for example, that on three recent occasions groups of escort vessels only arrived back on the Humber from Lerwick on the day they were due to sail out once more with the northward bound convoy. They had arrived at 9 a.m. and to keep to the schedule they had to sail again at 3.30 p.m. Such difficulties placed a great strain on crews, making it necessary to rush the completion of coaling and provisioning.[83]

Despite these problems, the institution of the convoy system along the North Sea coast can be considered a substantial success. Enemy U-boats, used to being able to pick off individual unarmed merchant ships, now found the seas to be virtually empty of such potential prey. To carry out their orders to sink mercantile shipping they had to find and then attack convoys in the war channels and shipping lanes, and these were usually defended by destroyers and an array of armed trawlers, and virtually every one of these craft was equipped with at least a modest supply of depth charges before the year was out. Many merchant ships voyaging alone had previously, of course, been sunk by U-boats which surfaced and finished them off by gunfire, but this was not an option when confronting convoys supported by several armed escort vessels. Torpedoes were now often the only option.

In addition, the U-boats now had to contend with hydrophone hunting drifter flotillas which listened for their approach to the shipping lanes. On the Yorkshire coast, for example, during August 1917, the Humber hydrophone drifter flotillas concentrated their efforts on waters adjacent to the war channel between Scarborough and Spurn. They formed a line four miles wide and after steaming for 15 minutes stopped engines and listened for a further ten before moving forward again.[84] Around the same date, further north, motor launches based on the Tyne maintained a night-time hydrophone listening vigil inside the war channel between Scarborough and Whitby.[85]

Nothing suspicious seems to have been picked up on these occasions but such tactics were continually varied thus adding a further dimension of risk and unpredictability for U-boats intent on wreaking havoc in the war channels. By now reports of submarine sightings were also often investigated by hydrophone listening activities along many stretches of the east coast. During that summer four Handley Page aircraft operating from Redcar also made what was considered a successful contribution to the coastal war, patrolling along inshore waters. Not all the aircraft being deployed on the North Sea coasts were British: on 3 September, a large American aircraft was reported to have bombed a submarine

83 NA, ADM 137/720, Auxiliary Patrol Humber, 6 December 1917.

84 NA, ADM 137/719, Auxiliary Patrol Humber, 11 August 1917.

85 NA, ADM 137/539, Auxiliary Patrol Tyne, 18 August 1917.

off Kilnsea.[86] Air support, whether by aeroplane or airship, was to play an increasingly important role during the last year of the conflict.

Nevertheless, the Germans continued their campaign and there were numerous reports of attacks on convoys during the year, which reached a peak in intensity during November. On 10 November, for example, a convoy of eight ships, escorted by five armed trawlers from the Tyne base, was attacked by a U-boat, which fired a torpedo that passed near the stern of the *John Gilman*, one of the new Admiralty Castle Class trawlers. The trawler straightaway sailed up the wake of the torpedo and dropped a depth charge but there was no further contact.[87] There was a further attack around the same time on a convoy south of Filey, but again without loss,[88] and in the middle of the month a substantial assault was mounted on a convoy of 17 ships being escorted by the Armed Yacht *Miranda*, six armed trawlers, and a division of Motor Launches. The *Miranda* responded by attempting a ramming run down the wake of the torpedo, but the sound of the passing convoy drowned out all attempts to locate the U-boat with hydrophones.[89] A further encounter, this time with a southbound convoy headed by the Hull armed trawler *Lord Reading*, occurred on 24 November. On this occasion, a periscope was sighted no more than 150 yards off the trawler's port bow. The *Lord Reading* immediately turned for ramming and crossed the submarine just forward of the periscope at a speed of 11 knots. A heavy grating was heard by all on board and a depth charge dropped as the periscope passed by. Owing to a heavy gale and strong seas the search was discontinued, but such actions must certainly have been at the very least unnerving for all on board the U-boat, which was probably damaged.[90]

A few days later, the *Lord Reading*, together with the armed trawler *William Browning* – another Admiralty class trawler which had come into service in 1917 – was involved in repelling yet another attack, this time by three U-boats in the war channel off Robin Hood's Bay. Enemy submarine attacks continued with great regularity along this stretch of coast into December, but with seemingly less success than was formerly the case.[91]

Shetland and the Convoys

Shetland had at first been made the transfer point for the Scandinavian convoys, because it offered the shortest crossing point to the Norwegian coast and minimised the risk of open sea attacks by U-boats. A U-boat attacked one of

86 NA, ADM 137/720, Auxiliary Patrol Humber, w/e 1 September 1917.
87 NA, ADM 137/2653, Auxiliary Patrol, monthly reports, December 1917.
88 NA, ADM 137/539, Auxiliary Patrol Tyne, w/e 10 November 1917.
89 Ibid., w/e 17 November 1917.
90 Ibid., w/e 24 November 1917.
91 Ibid., w/e 24 November 1917, w/e 15 December 1917, and w/e 15 December 1917.

the outward-bound convoys in early July, but was first fired on by the Armed Yacht *Amalthea* and armed whaler *Pilot Whale* before being depth-charged by the destroyer HMS *Arab*.[92] On 15 July, another potential attack on a convoy was probably thwarted when the armed trawler *Prefect* depth-charged a U-boat after spotting a periscope seven miles north-east of Sumburgh Head.[93]

During August, in accordance with Admiralty instructions, the Lerwick base organised several of its vessels into a special patrol to intercept and escort vessels from the White Sea and Murmansk coast, which were not all sailing by way of the Norwegian Fjords. The growing amounts of convoy traffic passing by way of the Shetlands along the Scandinavian routes continued to attract unwelcome attentions from the German vessels drawn to the archipelago, and on 18 August the Lerwick-based armed trawlers *Benjamin Stephenson* and *Elise* engaged with a submarine during which more than 40 rounds were exchanged. The U-boat had the bigger guns and stayed out of range for much of the 30-minute encounter, and although the submarine eventually disappeared the *Benjamin Stephenson* was badly damaged and taken in tow by the *Elise* before eventually sinking as they headed towards Bard Head.[94]

The levels of traffic led to other armed trawlers being transferred into Lerwick on a temporary basis in early October 1917. Thereafter, what became known as the Scapa Escort force was formed with the intention of convoying vessels between Lerwick and East coast ports over the following few weeks.[95] A week or so later the Germans changed their tactics and instigated a surface attack, along the lines of those that had taken place on Dover Patrol vessels. The Germans undoubtedly gathered evidence on the arrival and departure times of the convoys from neutral Norway, and shortly after daybreak on 17 October, a westbound convoy of a dozen assorted British, Belgian, Norwegian, Swedish, and Danish ships was attacked by two fast German minelaying cruisers, the *Brummer* and *Bremse*.[96] The convoy was being escorted by two Royal Navy destroyers, the *Mary Rose* and *Strongbow*, as well as the armed trawler *Elsie* and armed whaler *F. P. Fannon*, each deploying one 6-pounder gun. The destroyers were outgunned, and the *Mary Rose* soon sank with all hands, while the trawler *Elsie* gallantly stood by the heavily damaged *Strongbow* and took off many survivors before the stricken destroyer went down. The Cruisers also sank nine steamers before making off, but the *Elsie* and *F. P. Fannon* managed to escort the three remaining ships into Lerwick.

A week later, another westbound convoy was attacked, this time by a U-boat, and although two ships were torpedoed, one of them made it into

92 *Chatterton, Auxiliary Patrol*, 219.
93 NA, ADM 137/523, Orkney and Shetland, w/e 28 July 1917.
94 Ibid., w/e 18 August 1917.
95 Ibid., w/e 6 October 1917.
96 Chatterton, *Auxiliary Patrol*, 220–221.

Lerwick and was beached. But the Germans soon made another and this time very substantive surface raid on a Shetland convoy. The attack was well planned and came in two prongs. The attackers were two half flotillas of Germany's largest and fastest torpedo boat destroyers that had left Emden on 11 December, accompanied by the cruiser *Emden*. After reaching the Dogger Bank where the cruiser *Emden* remained, the two half flotillas split, one heading north and the other proceeding to position 25 miles north of Newcastle. The vessels that headed north had intended to catch a westbound convoy leaving Drammen, but were slowed by bad weather. They changed tack and eventually spotted an eastbound Scandinavian convoy, consisting of six ships and escorted by four armed trawlers, *Commander Fullerton*, *Livingstone*, *Tokio*, and *Lord Alverstone*, with two destroyers, the *Pellew* and *Partridge*. The British destroyers steamed off at speed, hoping to draw the Germans after them and away from the convoy. But while three of the German craft followed and engaged them in action, the fourth turned on the armed trawlers. The trawlers were hopelessly outgunned and the *Lord Alverstone*, *Tokio*, and *Commander Fullerton* were soon sunk.[97] In the meantime, the destroyer *Partridge*, disabled by a shell, also sank. Eventually, the whole convoy and escorts were wiped out except for the destroyer *Pellew*, which escaped in a shower of rain. Four officers and 48 survivors from the escorts were taken prisoner by the Germans, along with 23 civilians from the convoys.[98]

Meanwhile, the other raiders who had headed for the Northumbrian coast steamed northwards along the swept channel, then turned south again. They sank two steamers and a fishing vessel on this foray, before rejoining the *Emden* on the Dogger Bank and returning to Germany.[99]

The British response to these attacks was to change their arrangements, and in January 1918 Methil in Fife took over from Lerwick as the assembly port for the Scandinavian convoys.[100] To deter further surface attacks, regular squadrons of battleships and light cruisers were often at sea and met with the inward and outward convoys. After these changes, such German surface raids on the Scandinavian convoys came to an end.

During the three years from the beginning of 1915 to the end of 1917, the armed trawlers of the Auxiliary Patrol and the thousands of fishermen in the Trawler Reserve were almost continuously on the front line in this grim maritime war off the coast of the British Isles. For much of this time their armaments were woefully inadequate, but gradually, most notably in 1917, they had begun to benefit from improvements in wartime technology and new tactics, although that most crucial of weapons in the U-boat war, the depth charge, was still in

---

97 NA, ADM 137/523, Orkney and Shetland, w/e 15 December 1917.

98 Chatterton, *Auxiliary Patrol*, 222–224.

99 Ibid.

100 Marder, *From Dreadnought to Scapa Flow*, 4:314.

relatively short supply. The ranks of hired trawlers had also been strengthened by the acquisition of high-speed launches and by late 1917 were being supplemented by the Admiralty class trawlers and drifters. But, perhaps above all, years of active service and effective training had forged these fishermen and their vessels into powerful fighting units, which were to be of crucial importance as the Great War entered its final and ultimately decisive year.

Chapter Five

Fighting Overseas

Dardanelles, Gallipoli, and the Eastern Mediterranean

The international conflagration that the Great War became soon carried fishermen to conflicts on coasts far removed from the shores of the British Isles: not least, and from quite early on, the waters of the eastern Mediterranean. In January 1915, as British and French plans for a naval assault on the Dardanelles were being formulated following Turkey's entry into the conflict on the side of Germany and Austria-Hungary in late October 1914, an initial flotilla of 13 Auxiliary Patrol trawlers was sent out from England. By 21 February, 21 trawlers had reached Malta, and some of the vessels with their fishermen crews had already embarked on the next stage of the voyage to the Dardanelles.[1]

The initial allied strategy involved forcing a passage through the Dardanelles, the narrow straits that divide Europe from Asia, by reducing Turkish defences at the entrance to the passage. The trawler minesweepers were required to clear a way for large warships to move in and bombard the enemy's coastal positions, the outer defences of which included two forts on the Dardanelles Peninsula and two others on the Asiatic side, as well as several minefields. The whole strategy depended heavily on the creation of a swept and marked channel for the warships to close in.[2]

The first trawlers arrived and commenced sweeping on 19 February, working to within 5,000 yards of Gaba Tepe, and they managed to buoy a channel after encountering no mines. Bad weather then prevented operations for nearly a week, but on the night of the 25th the minesweepers renewed their efforts, sweeping the entrance to the Straits, cover being provided by a couple of battleships and a number of destroyers. Once again, no mines were encountered, and the battleships entered the Straits the next day and began bombarding the forts. After dusk fell, the trawlers entered the Straits once more and made a four-mile sweep, but again found no mines. Then bad weather disrupted the task once more; but on the moonlit evening of 1 March, the trawlers renewed their efforts, managing

[1] Hurd, *History of the Great War*, 2:158–159.

[2] Patrick Gariepy, *Gardens of Hell: Battles of the Gallipoli Campaign* (Lincoln, Nebr., 2014), 17.

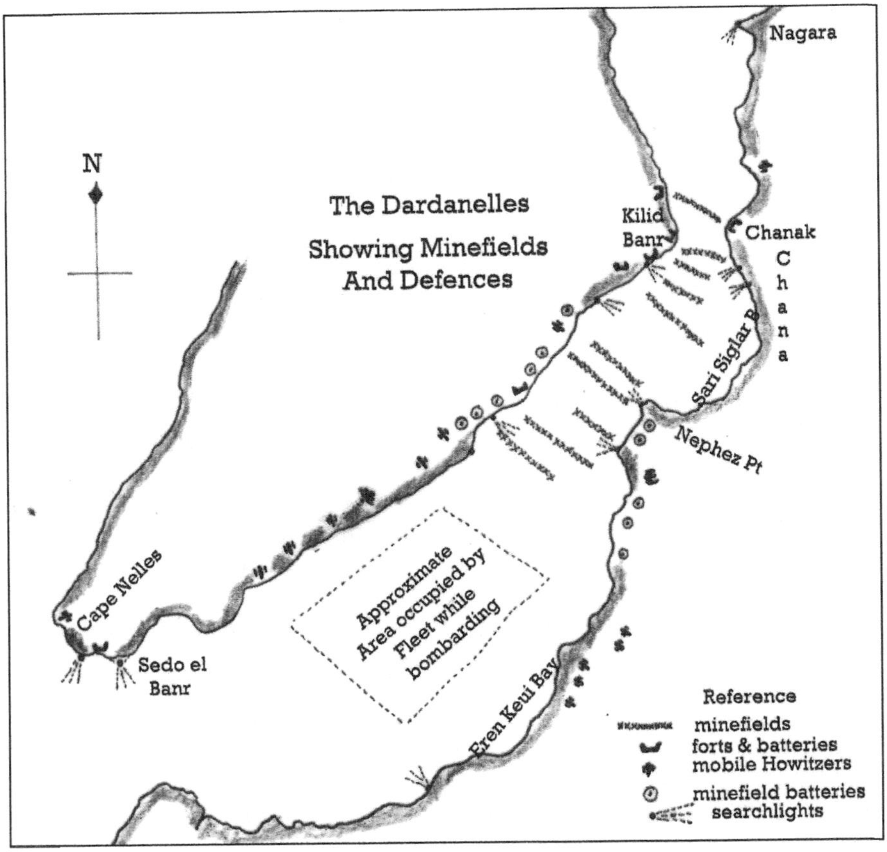

Figure 7: Outline map of the Dardanelles
Source: Ken Knox

to sweep to within 3,000 yards of Cape Kephez. They came under fire from Turkish batteries once abreast of the Suandere River and had to withdraw under the protection of a smokescreen made by accompanying destroyers.[3]

During the following day, the battleships restarted their bombardment and that evening the trawlers returned and continued their work amongst the Kephez minefield, but came under very heavy fire from enemy guns and made no further progress. On the evening of 6–7 March, they tried yet again to clear a passage through this minefield, but were picked up by searchlights on the shore, and came under further intense fire from coastal batteries in the Narrows (Figure 7).[4]

3 Hurd, *History of the Great War*, 2:159–161.

4 Ibid.

The majority of trawlermen involved in these actions had still been fishing the previous year and since joining the armed forces had never tried to sweep up mines during the hours of darkness or attempted to carry out such operations in a narrow stretch of water flanked by searchlights and enemy artillery. Theirs was a seemingly impossible task and the allies had reached an impasse: the battleships required a channel sweeping through the mines so they could close and destroy the enemy gun batteries, but the trawlers were unable to sweep the mines until the enemy guns were silenced.[5]

The trawlers, this time accompanied by picket boats, returned to the task on the night of 8–9 March, but success eluded them, and the hired Grimsby trawler *Okino* struck a mine and only the skipper and four crew survived.[6] The allies redoubled their efforts, sending in seven trawlers and a couple of picket boats with explosive sweeps in yet another abortive attempt the following night.[7]

The strong 3- to 4-knot current which ran in the direction of the Mediterranean off Kephez complicated matters. It was difficult for trawlers sweeping against this flow to make much headway and on the following night they tried a somewhat different approach. The new plan involved steaming to the north, to the other side of the minefield, then sweeping back down with the current. But almost as soon as the sweeps were deployed by the first pair of trawlers they ensnared two mines which detonated in the dark waters, sinking the trawler *Manx Hero*. All the crew were rescued by Skipper Woodgate on the trawler *Koorah*, but the element of surprise was gone.[8] The Turkish forces had found the minesweepers with their searchlights and the fishermen were once more subjected to heavy fire. The battleship HMS *Canopus* tried to bombard the searchlight positions from a distance, but without much success. Two of the other minesweepers were hit by Turkish fire and the crew of a third pair of trawlers were much too disconcerted by the heavy gunfire to organise an effective sweep and so yet again the minesweepers had to withdraw under cover from the destroyers.[9]

Another attempt at a sweep from the north was made, but the Turks were now wise to the tactic and this time when the searchlights illuminated the vessels and the Turkish guns on the shore opened up the lead trawler turned and, followed by the others, steamed back down the Dardanelles. This refusal has since been a source of controversy: up to this point the trawler crews had done everything that was required and, as Admiral Roebuck reported, 'in some cases the crews appear to have no objection to being blown up by mines though they do not like working under gunfire'.[10] After this, three Royal Navy sailors were added to the trawler

---

5 Marder, *From Dreadnought to Scapa Flow*, 2:71.

6 Corbett, *Naval Operations*, 2:199.

7 Ibid., 205.

8 Dorling, *Swept Channels*, 147.

9 Ibid.

10 Corbett, *Naval Operations*, 2:207.

crews and all were volunteers.[11] The following night French minesweepers also tried their hand but found they could not steam against the current.[12]

A further foray was carried out on 13 March, after HMS *Cornwallis* spent an hour bombarding the offending lights and batteries. The trawlers were now supported by picket boats and escorted by HMS *Amethyst* and several destroyers. But the flotilla was again spotted making its way upstream, searchlights were turned on, and the salvoes from the shore resumed. When close to the Narrows, the point where the waterway is no more than 300 metres wide, the trawlers came under withering fire as they gathered together while trying to deploy their sweeps. The Grimsby Trawler *Fentonian* and the *Star of the Empire* from Lowestoft collided during the action, and three crew, two of whom were from the Trawler Reserve, lost their lives.[13] The *Amethyst* was also badly damaged, collecting many casualties when it exposed itself to the Turkish artillery in an attempt to draw fire from the trawlers. During these attempts to clear the Narrows, the trawler winches and wheelhouses were often badly damaged by enemy fire, and the later reinforcements had armour plating fitted as additional protection when they called into Malta on their voyage to the eastern Mediterranean.[14]

The numbers killed or wounded and the artillery damage to trawler decks and winches prevented most of the vessels from deploying their sweeps on this attempt.[15] One pair of vessels did manage, against the odds, to start operations. A few mines were cleared but the cost was extremely high. Almost everyone on board two of the trawlers was either killed or wounded, and in all, four trawlers and a couple of picket boats were also badly damaged. Twenty-seven men were killed and 40 wounded in this one action.[16]

Yet another tactic was then tried: several destroyers were fitted out for minesweeping and on 18 April two of them swept ahead of the bombarding warships *Triumph* and *Majestic*.[17] Such fast sweepers might have made a greater difference to operations, but they were not available in sufficient numbers. Although a combination of Royal Navy and French warships did manage to inflict some damage on Turkish forts around the Narrows during a subsequent attack, the action was far from an overall success and several warships succumbed to mines, none of which was swept. It was becoming very apparent that plans to seize the waterway would not succeed by using warships alone: a military landing and the occupation of the Gallipoli Peninsula seemed the only alternative available.

---

11 Gariepy, *Gardens of Hell*, 19.

12 Corbett, *Naval Operations*, 2:193.

13 Gariepy, *Gardens of Hell*, 20. See also Commonwealth War Graves Commission Records, Gilbert Morvinson, William Isaac Barnes and Edward Timothy Worrall.

14 Hurd, *History of the Great War*, 2:162.

15 Dorling, *Swept Channels*, 150.

16 Chatterton, *Auxiliary Patrol*, 74–75.

17 NA, ADM 186/604, 'History of British Minesweeping in the War', 25.

By the end of April 1915, the efforts of the destroyers had created a swept area that reached to within 8,000 yards of the Narrows forts and seven mines had been removed in the process.[18] But the shore-based guns opened up whenever the destroyers started sweeping. Although these warships sustained fewer casualties than the trawlers, the strain on the crews was considerable and it was finally decided to abandon all further attempts to sweep inside the straits until landings of troops had altered the equation. In all, there had been 12 attempts to sweep the Kephez mines at the entrance to the Narrows of which only two had succeeded in penetrating the minefield.[19]

In retrospect, the conditions in which the minesweepers had to work were the causes of their lack of success. Not only were the crews themselves relatively new to the task (most had never previously been under fire), but they were also required to carry out their work at night, something that had not really been attempted before and was hardly ever tried again during the rest of the war.[20] The trawler crews were not civilians, as has been suggested by some commentators. Many were in the Trawler Reserve, but most had received only limited training by this point and had little or no experience of minesweeping under fire. The strong currents meant that it was difficult for the trawlers to make much headway against the flow. The later attempts to sweep with the flow meant that the trawlers had first to voyage to the other end of the minefield and thus steam through heavy fire before they even reached the position where they could deploy their gear. Indeed, during all these sweeping operations, the withering bombardment from both sides of the Straits had been a crucial factor in their failure and one that the warships that accompanied the minesweepers had been unable to neutralise.

It had also been evident from early in the campaign that the harbour at the strategically important port of Smyrna would need to be put out of action to prevent it becoming a base for any U-boats which were sent to support the Turks. Smyrna was first bombarded in early March. But here also a channel had to be cleared for the bombarding warships, and once again the trawlers came under attack from Turkish shore-based batteries soon after they commenced sweeping.[21] The attack here was eventually called off when the large warships were recalled to the Dardanelles, but as it turned out Smyrna was not afterwards much use for U-boats because the Turks had blocked it with sunken ships to thwart an allied attack. Minesweeping operations that continued for some time close to the entrance to the Dardanelles were also subject to enemy bombardment from the shore and were eventually discontinued.[22]

18 Ibid.

19 Ibid., 24.

20 Ibid., 25.

21 Corbett, *Naval Operations*, 2:197–198.

22 NA, ADM 186/604, 'History of British Minesweeping in the War', 25.

Photograph 7: War-time Auxiliary Patrol crew of the Buckie drifter *Obtain*
Source: Jon Grobler

During a period of comparative quiescence while preparations for the Gallipoli landings were carried out, the minesweeping forces carried out further training. On 25 April, the army embarked on its attempt to occupy the Gallipoli Peninsula and trawlers played a key part in the amphibious landings. Fourteen trawlers in all were used during these landings. Four of these transferred the King's Own Scottish Borderers from transport ships to Y Beach where they grounded, and the men then made the shore by boat. They were followed by four other such vessels full of Royal Marines. A couple of other trawlers assisted in the landings at W Beach while the South Wales Borderers were landed in cutters from four trawlers convoyed from the battleship HMS *Cornwallis.* A further six trawlers were also deployed on minesweeping duties in Xeros Bay on the night of 24–25 April where a feint of landing was made.[23]

The trawlers and crews were fully engaged for much of the remainder of the ill-fated Gallipoli Campaign. July was an exceptionally strenuous month for these vessels which were kept extremely busy on tasks which included towing barges, ferrying new troops and casualties, or conveying supplies of ammunition to the shore after dusk. The trawler crews got few chances for breaks while coping with oppressive heat and foul infestations of flies, which spoilt their food before it even left the galley.

23 Chatterton, *Auxiliary Patrol*, 78.

Reinforcements continued to arrive from Britain to enlarge this auxiliary fleet, and by the end of July Admiral Roebuck had 47 minesweeping trawlers available, together with 31 net drifters and 21 other armed trawlers, as well as a number of faster fleet sweepers and a few motor boats. Trawlers also assisted in the blockade of Smyrna and every spare vessel stationed in the Aegean Sea area was needed for anti-submarine patrol work. Auxiliary Patrol vessels were also engaged in the actions surrounding the landings at Suvla Bay, the last major amphibious landing of the ill-fated campaign. This began on 6 August, and on the 16th, two Hull trawlers, the *Lundy* and *Poonah*, were sunk after collisions in the bay.[24]

The trawlers had faced a baptism of fire since arriving in the eastern Mediterranean and eventually – in July 1916 – a couple of Hull skippers of Grimsby trawlers, Alfred Swain of the *Escallonia* and Alfred Berry of the *Frascati*, were awarded the DSC for their services.[25] Both had been in action on many occasions and had carried out their orders even when constantly under fire. Both were also to have a long and interesting war.

More reinforcements flowed in during the autumn and many were utilised in actions around Salonika and found employment protecting transport ships. Auxiliary Patrol vessels also played crucial roles in the final evacuations of Suvla and Anzac on 19–20 December and Hurd tells us that over 42,000 troops were taken off by the trawlers and fleet sweepers while net drifters provided protection for the monitors and battleships that bombarded the Turkish positions.[26] Trawlers also came under heavy fire while playing a full part in the final evacuation of Cape Helles on 7 and 8 January 1916.

The last evacuations at Gallipoli did not, by any means, bring an end to Auxiliary Patrol activities in the eastern Mediterranean. Extensive minesweeping operations were maintained around the base at Mudros on the island of Lemnos, and patrols continued between Malta and the Aegean and around the Greek islands as the war progressed. Like their colleagues operating off the British coasts, the hired fishing vessels deployed in these waters gradually became better armed, eventually carrying depth charges, larger guns, and hydrophones.[27]

The Gallipoli Campaign played a significant part in encouraging the Germans to extend their U-boat and mining campaign to Mediterranean waters. On the very day that the allied troops began landing on the Gallipoli beaches, Otto Hersing, the ace German U-boat commander, embarked on a voyage from the River Ems which took him by way of the Fair Isle Channel and the Straits of Gibraltar to the Adriatic Sea. Cattaro (now Kotor in Albania) and Pola (now

[24] *British Vessels Lost at Sea, 1914–1918*, 12.

[25] S. J. Eyre, *Skippers of the Royal Naval Reserve in the First World War* (Leamington Spa, 2014), 176–177 and 572–573.

[26] Hurd, *History of the Great War*, 2:164–165.

[27] Ibid., 1:164–165.

Pula in Croatia) were then major naval ports of the Austro-Hungarian Navy, and became bases for German U-boats for the remainder of the war. Hersing reached Cattaro on 13 May 1915, but, despite his epic voyage, he did not linger long and left a week later on a cruise which took him much further eastwards where he sunk the pre-dreadnought battleships *Triumph* and *Majestic* before sailing up the Dardanelles and docking at Constantinople on 5 June. Shortly after leaving Constantinople in early July, he torpedoed the French liner *Carthage* then being used as a troop transport when it was lying at anchor off Cape Helles, Gallipoli.[28]

The Straits of Otranto

Germany, regarding the Gallipoli Campaign as of crucial strategic importance, soon sent several U-boats to operate from Constantinople, and by the autumn of 1915 had also built up a significant U-boat presence at both Cattaro and Pola. The Austro-Hungarian Navy had its own submarines, but the growing German U-boat presence posed a serious additional danger to allied operations in the Mediterranean, threatening lines of communications with Gallipoli. Even before the full force of the enemy reinforcements was felt in the Mediterranean, it was decided to try to block the exit from the Adriatic through the Straits of Otranto by deploying nets in a somewhat similar fashion to those being used in the Dover Straits. This was now potentially possible because Italy had entered the war on the side of the allies on 24 May 1915. Laying nets here was to prove quite different in many ways, for while the waters off Dover were comparatively shallow, the Straits of Otranto were between 300 and 500 fathoms in depth and although the tides were not a factor to contend with here the defile to be barred was almost twice as wide (Figure 8).[29]

To create this new barrier, 60 Auxiliary Patrol drifters and their crews were despatched from England in September 1915, and the number almost trebled during the following year. At the end of the month the drifters started deploying their nets, which could only cover sections of the surface waters of the Strait. Although these fishing vessels faced a formidable task in trying to maintain a barrier of any sorts by towing their nets across this deep and wide divide, they did succeed in enmeshing two U-boats during the first winter of operations; but many encounters with U-boats were not to their advantage. On 12 October, the drifter *Restore*, which, other than its nets, possessed nothing more than five rifles in the way of armaments, was shelled after encountering the U-39 on the barrier. The U-boat turned both its guns on the drifter, and after disabling the engine room closed in to finish the job. Three more drifters were also attacked around

---

28 Gibson and Prendergast, *German Submarine War*, 71–72 and 74.

29 Ibid., 124.

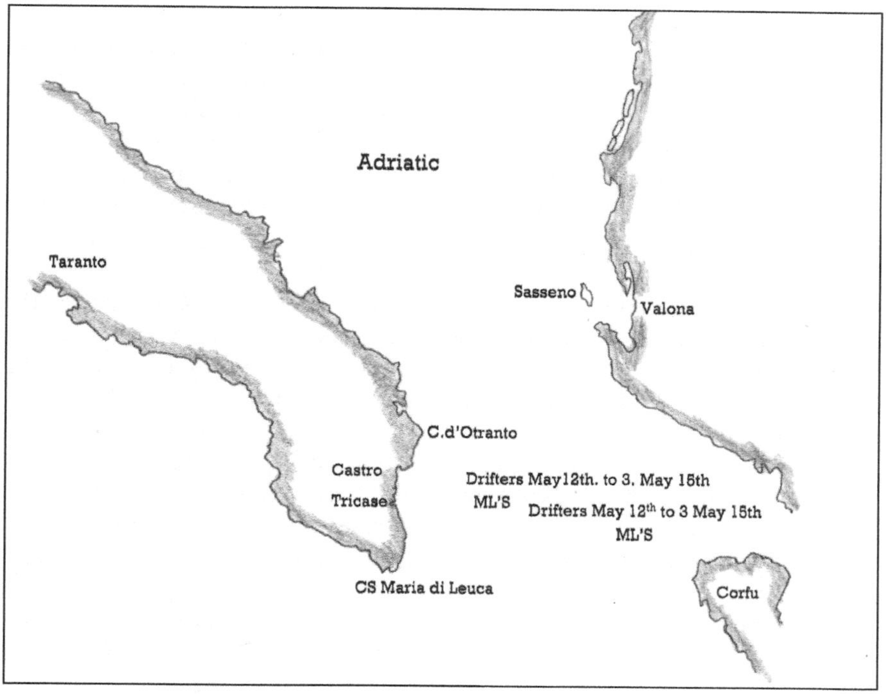

Figure 8: Outline map of the Straits of Otranto
Source: Ken Knox

the same time though without loss. By mid-November 1915, all the drifters had been armed.[30]

From the beginning of December, many of the drifters working on the barrier were diverted to the task of assisting with attempts to evacuate the remnants of the Serbian army, which was retreating into Albania after a series of crushing defeats. Thirteen of these drifters, participating in the later stages of the evacuation of Serbian troops from the Albanian port of Durazzo (better known today as Durres), came under fire from guns placed in the surrounding hills and from attacking aircraft.[31] The vessels were also fully involved in many other dimensions of this rather unknown element of the war: seven of the drifter skippers were later awarded the DSC for their part in these actions.[32] Not only were many soldiers and refugees carried to safety on these erstwhile fishing vessels, but others covered the landings by Italian troops at Valona (now Vlore

30 Paul Halpern, *The Battle of the Otranto Straits* (Bloomington, Ind., 2004), 23.
31 Chatterton, *Auxiliary Patrol*, 127.
32 Ibid.

in Albania), while yet more drifters protected steamers taking supplies to the hard-pressed Serbian and Montenegrin forces on shore and picked up troops and sailors from sinking vessels. Back on 4 December 1915, for example, the drifters *Evening Star* and *Lottie Leask* had closed in and rescued 500 Italian troops from a transport ship that struck a mine off Cape Linguetta, only just managing to chop away the ropes that the soldiers had slipped down before the ship plunged to the bottom of the Adriatic. The same day, the drifter *Manzanita* rescued the crew of the Italian destroyer *Intrepido*, which had been fatally damaged by a mine laid by UC-14 off Valona.[33]

However, after participating in this rescue the *Lottie Leask* did not see the month out, being attacked by two submarines just a few hours after leaving the Italian port of Brindisi. She was abandoned after being hit by five shells. The drifter's crew made it to the Adriatic coast in their rowing boat, and although they landed in territory seemingly held by the Austrians they eventually made their way to Valona, completing the final landward stage of their journey with some Serbian troops, where they were shipped back to Italy.[34]

Four Auxiliary Patrol drifters had been lost by the time the Durazzo evacuations were completed at the end of February, and although they could now focus on the Otranto Straits net barrage, two more of these craft, the *Boy Harold* and *Enterprise* II, were sunk by mines. It was also soon all too clear from attacks on allied shipping elsewhere in the Mediterranean that Adriatic-based U-boats were still passing through the barrier, and in April the depth of the nets was increased to 140 feet and the tops were submerged a further 12 feet below the surface. Even with this additional depth, however, it was still possible for a daring U-boat commander to dive below them, although he risked damaging or even losing his craft.[35]

Despite their barrier's shortcomings, the drifter crews soon reaped some reward for their attention to this seemingly thankless task. On 9 May, the Austrian submarine U-6 started on a cruise from Catarro and travelled on the surface until it dived on seeing one of the Otranto drifters, but fouled a propeller in the nets. The U-boat resurfaced to cut away this entanglement, but its movement of the nets had been detected by the drifter *Calistoga* and it was then seen when surfacing by both the *Dulcie Doris* and *Evening Star* II. The *Evening Star* II slipped its nets and opened fire, seeming to hit the Austrian vessel. The U-boat commander opened the sea valves and ordered his crew to abandon ship. The submariners were picked up by the drifters and carried in triumph into Brindisi. The drifter crews were later rewarded with £1,000 by the Admiralty.[36]

---

33 Hurd, *History of the Great War*, 2:167.

34 Chatterton, *Auxiliary Patrol*, 127–129.

35 Tim Benbow, *Naval Warfare 1914–1918: From Coronel to the Atlantic and Zeebrugge* (London, 2008), 198.

36 Hurd, *History of the Great War*, 2:169–170.

On 7 July, a northward-bound U-boat also got entangled in the nets and managed to escape, but this appears to have provoked a retaliatory attack by the Austrians a couple of days later, which succeeded in destroying the Banff drifter *Astrum Spei*[37]and the *Clavis* of Buckie, and inflicting damage on two others. Nine crew from the drifters were taken prisoner, one of whom, 17-year-old trimmer Walter Watt, jumped over the side from the Austrian cruiser. He was soon recaptured but leapt overboard again, this time getting away and after 90 minutes in the sea he was picked by a British drifter.[38]

This raid illustrated the exposed and relatively unprotected nature of work on the Otranto Barrage for the crews of the fishing vessels which worked the nets. Indeed, the casualty toll might have been higher had not the cruiser broken off its assault as a couple of Italian destroyers approached. Even so, this Austrian attempt to render the barrage unusable did suggest that, for all its shortcomings, the enemy perceived it as something of a threat to their submarines.[39]

It is possible that the drifters exacted some degree of revenge within the month when the *Quarry Knowle* noticed signs that indicated a submarine had fouled the nets and summoned the *Garrigill*. Both craft saw the net was sinking and its cables tightening. After the *Garrigill* dropped a depth charge there was a huge underwater explosion. The drifter dropped a second device, but the strain on the cables did not ease so the *Quarry Knowle* began towing the heavy net towards shallower water. The *Garrigill* tried to assist, shackling a wire to the gear, but in vain, as the cables eventually parted, and the nets sank, and the drifter crews thought the submarine had gone down with them.[40] It is possible that this vessel was the UB-44 which had sailed from Cattaro for Constantinople on 4 August 1916 and was not seen again: its wreckage has never been located.[41]

Whatever was the cause of its demise, this loss of a U-boat seemingly provoked yet another enemy attack on the barrage. On 26 August, three Austrian aircraft flew in from the north to attack a drifter group on the line. Two went for the drifter *Rosie*, which was sunk when a bomb hit the fore hatch, while the third aeroplane machine-gunned the drifter *Craigbo*. Soon all the planes had turned their guns on the *Craigbo*, but the other drifters closed in around their target, and whenever the aircraft came in low enough to be targets they opened fire with their guns. The aircraft regrouped and then renewed their attack on the *Craigbo*, this time with bombs, but they missed their target. Eventually, the aircraft broke off and flew off northwards. After this attack a programme of

---

37 *Aberdeen Press and Journal*, 15 July 1916.

38 Watt was later advanced to the rank of engineman and awarded the Conspicuous Gallantry Medal. See Halpern, *Otranto Straits*, 120.

39 Ibid., 131.

40 Chatterton, *Auxiliary Patrol*, 158.

41 Halpern, *Otranto Straits*, 38.

installing anti-aircraft guns on the drifters was commenced and the following spring a seaplane base was established to guard the Otranto Straits.[42]

During August 1916, the first of the new motor launches, the MLs as they were known, reached the Straits, followed a few weeks afterwards by the steam yacht *Catania*, which assumed the role of their mother ship. They operated out of the Italian harbour called Gallipoli, and a few days later the drifter base was transferred from the port of Brindisi to Taranto. To try to counter the vulnerability of the barrage, the position of the net line was shifted southwards. In September, Admiral Kerr established a drifter line which stretched between Cape Santa Maria di Leuca on the heel of Italy across to Fano Island on the Albanian coast.[43]

The work of the drifter patrol continued throughout the remainder of 1916 and throughout 1917. On 17 December 1916, yet another U-boat entanglement with the nets led to an action involving a number of drifters including the *Fisher Girl*. Although the Austrian submarine U-20 was believed to have been sunk in the action, this was in fact not the case as this craft remained active until it was finally lost in July 1918, and no actual submarine seems to have gone missing at the time of this drifter action. The U-boat involved no doubt escaped, but the incident probably provoked the next attack on the barrage by Austrian forces, this time on 22 December. On this occasion, a light cruiser and four destroyers ran down the Adriatic and shelled a couple of drifters, the *Gowan Lea* and *Our Allies*, both of which were badly damaged. The Austrians then made off when sighted by six French destroyers. Fortunately, all the drifter crews survived the attack.[44]

On 15 May 1917, the Otranto drifters were attacked in force once more. The attack was devised by Nikolaus Horthy, commander of the light cruiser *Novara*, who planned to make what he termed a 'clean sweep' of the barrage.[45] Seven divisions of drifters were still at their nets when three Austrian cruisers made their surprise attack as dawn was breaking. One started shelling the drifters from the western end of the blockade while the other two commenced their actions from the eastwards. Given the disparity in armaments, the cruisers were often able to stop to open fire and sometimes took prisoners, providing time for the crews to abandon ship. The attack lasted for three-quarters of an hour and despite the overwhelming superiority of the Austrians a few drifters did return fire, most famously the Fraserburgh Skipper Joseph Watt and his crew of eight sailors and a dog aboard *Gowan Lea*, which had by now returned to service after being damaged in the earlier attack. Watt was confronted by the cruiser *Heligoland* which demanded the drifter surrender, but he ordered the crew to open fire with their six-inch gun. Other drifters also joined in, but the *Gowan Lea* was badly

42 Chatterton, *Auxiliary Patrol*, 158.

43 Laurence Sondhaus, *German Submarine Warfare in World War I: The Onset of Total War at Sea* (London, 2017), 80.

44 Chatterton, *Auxiliary Patrol*, 170.

45 Halpern, *Otranto Straits*, 44.

damaged in the exchange. The Austrian vessels made off, and despite the damage suffered by his drifter Watt and his crew are said to have rescued men from other sinking vessels and provided medical aid. Fourteen drifters were sunk in the action causing a large part of the barrage to be taken out of action for some time. The Austrian forces were later that day engaged by Italian and British vessels in what became known as the Battle of the Otranto Straits, which was the largest Mediterranean area naval engagement during the Great War.[46]

Afterwards, many of the drifter crews were complimented on their courage in the face of such force, but there was private criticism at the time of some of those drifter crews who were said to have surrendered without a fight even though the odds were clearly stacked against these lightly armed barrage craft. Watt was awarded the VC for his action, one of two fishermen to receive this highest of decorations during the Great War.[47] Several others involved received the Conspicuous Gallantry Medal or Distinguished Service Medal. Six skippers, including Watt and also an engineer, hailing from the ports of Banff, Buckie, Fraserburgh, and Peterhead, were also awarded silver and bronze medals by the Italian king.[48]

After this attack the drifter line was moved further south and the vessels began working only during daylight hours, taking shelter each evening at places along the coast which allowed the U-boats virtual freedom of movement by night, although during periods of submarine alert the nets were sometimes deployed by the drifters after dark, albeit ostensibly under the protection of larger warships when these were available. Some senior officers certainly considered the barrier to be completely ineffective at this juncture, and the lightly armed drifters needed stronger support in the form of a combination of well-armed trawlers as well as aircraft and destroyer patrols. Indeed, although more modern analyses of U-boat activity suggest that the Adriatic drifter barrages were usually more of an irritation than a barrier to submarines, the fact that they were subject to attacks suggests that the enemy as well as sections of the allied commands thought they were more effective than they were.[49]

By this stage, the barrage drifters and crews were exhibiting signs of wear and tear. In October 1917, it was decided that skippers and crews who had been on station for more than two years should be relieved, and returned to service in home waters when replacements could be shipped out, although this took some time to expedite. On return to Britain they were granted a one month period of leave before being assigned to other duties.[50] The drifters with wooden hulls were also suffering from their sojourn in the southern Adriatic seas. In the early

46 For a full account, see ibid.

47 Ibid., 117–121.

48 *Aberdeen Weekly Journal*, 23 November 1917.

49 Halpern, *Otranto Straits*, 149.

50 NA, ADM 137/1579, Otranto Barrage, 23 March 1918.

days of the barrage they had received very little attention and consequently many had been attacked by worm. Later a quick slipping and docking routine was adopted, and every six months the wooden drifters were scraped and coated with a mixture of vegetable tar (65 per cent), coal tar (15 per cent), and sulphur (20 per cent), which was recommended by Italian naval authorities, and new planks were fitted where necessary.[51] Though such actions seemed to retard the entry of new worms they did not seem to do much for removing the existing infestations, and it was decided to return the wooden drifters to home waters and send out steel hulled versions as replacements. Though it took many months to collect sufficient steel drifters from various patrol areas around the British Isles, the wooden ones were eventually replaced and sent back to resume duties in home waters. Amongst the wooden vessels which eventually went back under this programme was the *Gowan Lea*, of Ottranto Barrage fame. Many of the drifters which arrived were fitted with hydrophones which were to play a major role during the final year of the war when the laying of an ambitious fixed net barrage across the Straits was embarked upon, and will be discussed in more detail in Chapter 7.

Auxiliary Patrol vessels and crews were by this time also found in many parts of the Mediterranean. Six trawlers had been sent to Alexandria before the end of 1915 and they were later joined by others, their job to patrol shipping lanes from Crete to Port Said. Meanwhile, armed trawlers conveyed troops along the Egyptian city of Mersa Matruh, to the west of Alexandria, and carried out anti-submarine patrols along the coast. Maltese harbours were mined for the first time in the last week of 1915, and Auxiliary Patrol vessels had also to counter enemy submarine minelaying activities, which were fairly extensive in 1916, covering waters off the coasts of Genoa, Marseilles, Taranto, Brindisi, Crete, Mudros, Alexandria, and Malta, amongst other places.[52]

The White Sea and the Northern Flotilla

But war service also took fishermen and trawlers far to the northwards of home waters. In June 1915, the German auxiliary cruiser *Meteor* embarked on a minelaying voyage north of Russia, laying a substantial number of mines in the seaways to Archangel, the intention being to disrupt vital supplies being shipped to Russia. Archangel was then of crucial strategic importance when supplying Russia with munitions and other goods, as the sea routes through the Black Sea and Baltic were by then in enemy hands, and the new railway to what became the ice-free port of Murmansk was not completed until 1917.[53]

---

51 Ibid., 20 December 1917.

52 NA, ADM 186/604, 'History of British Minesweeping in the War', 41.

53 Gibson and Prendergast, *German Submarine War*, 111.

The steamer *Arndale* was the first casualty of the mines, losing three crew when blown up on 11 June 1915, and nine other merchant ships of varying nationalities were also subsequently either lost or damaged. A flotilla of six trawlers and a couple of store ships were soon fitted out at Lowestoft and sailed on 22 June under the charge of Acting Commander Leopold A. Bernays, RN, who had previously been involved in sweeping the Scarborough and Tory Island minefields. Each trawler was armed with a 12-pounder gun and the lead trawler was fitted with a wireless.[54]

The flotilla voyaged northwards by way of Lerwick and reached Alexandrovsk along the Murmansk coast on 6 July. Within a few days of arrival the vessels had swept up several mines, although one device damaged the Hull trawler *T. R. Ferens* when it detonated. Within another week, 50 mines had been cleared from the seaways and the search continued, all the while the vessels and crews having to contend with strong tides, gales, snow, and ice while working on this largely unlit, unbuoyed, and starkly unwelcoming coast which offered little in the way of shelter or support. During August, two more minesweeping trawlers and a collier arrived to assist, and by October 150 mines had been removed, mainly by the British trawlers, although a few were also dealt with by the Russians.[55]

The original trawler force voyaged back to Britain to refit in December 1915, but was practically doubled in strength the following year. By 6 June 1916, ice conditions had improved sufficiently for vessels to reach Archangel that year without the help of ice breakers and the minefields between Svyatoi Nos and Sozonova, which had been discovered the previous year, were reswept and several stray mines subsequently located. In early July, a field evidently laid by an enemy submarine was located, and by mid-August 26 of these mines had been dealt with, but not without loss. During these operations, the minesweeping trawler *John High* was blown up on a mine, and only one member of the crew was pulled from the heavy ice-cold seas alive. A Russian steamer and schooner were also destroyed before the minefield was rendered safe.[56]

Locating and destroying mines in these harsh waters presented its own challenges for the trawlers and their crews. Being so far from base they had little in the way of an engineering infrastructure to support them. There was no dry dock to repair damage, and rough and ready repairs to propellers and hulls were often made by beaching the ships, and thus running the risk of being caught in an exposed position by sudden storms blowing in from the seaward.

The minesweepers stuck to their difficult duties throughout the summer and autumn of 1916. On 20 September, the British steamship *Elton* struck a mine, although in the buoyed and seemingly swept channel that had been created, and a further three steamers were sunk in October. Work was reorganised, and

---

54 Corbett, *Naval Operations*, 3:49.

55 Hurd, *History of the Great War*, 2:68–69.

56 Chatterton, *Auxiliary Patrol*, 156.

ten more mines were located and destroyed, but the need for vigilance was unremitting. Fresh mines, further to the seaward, were found after a Russian steamer fell victim in November, and by the time some of the trawlers started to return to Britain for a refit in late 1916 they had destroyed 42 mines. It later emerged that German submarines had laid 72 mines in these waters during the year, which, allowing for losses and mines which would have broken adrift in rough seas, this was a good sweeping record in such difficult conditions. Half-a-dozen minesweeping trawlers remained in Russian waters until the end of the year and beyond, and were certainly kept busy on sweeping, patrol, and boom duties. On 20 December 1916, HM trawler *Lanercost* (which in peacetime worked out of Fleetwood) together with the *Holyrood* and *Arctic Prince* – respectively from Grimsby and Shields – refloated a French vessel, the SS *Consul Horn*, which had gone ashore to the east of Letenski Point, and brought it safely into the Kola Inlet.[57]

These northern sea lanes which the trawlers patrolled and swept were immensely important for the Russian war effort. In 1915, some 700,000 tons of coal and another 500,000 tons of general cargo, mainly war materials, had voyaged by this route, and in 1916 that figure doubled to 2.5 million tons, which was more than 25 times the normal traffic levels in the White Sea (Figure 9).[58] They were to remain of key importance down to the Bolshevik Revolution.

These trawlers had only been able to return home for a refit after being relieved by another flotilla under the charge of Commander Evelyn Gore Langton, RN, which had left Yarmouth Roads on 1 January 1917 and arrived, by way of Lerwick, at Murmansk, their base in the Kola Inlet, a fortnight later.[59] Langton's flotilla then remained on station throughout the worst of the winter, the weather characterised by high winds at sea, occasionally interspersed with a few hours of calm weather and frequent snowstorms which made navigation difficult, especially at night. Dense fogs were also on occasions a major hazard. Twenty or 30 degrees of frost were reported to be usual temperatures, and as low as 52 degrees had been registered, The cables and charges on the modified sweeps fitted to some of the trawlers were reported at one time to be continually frozen in a solid mass of ice and impossible to chop out without damaging the circuit or running the danger of exploding the charges.[60] It was not until 25 March that the temperature rose above freezing point during the day, and the Kola Inlet was frequently covered with ice, though this was not usually thick enough to impede shipping movements, and was normally broken up by spring tides.[61]

---

57 NA, ADM 137/1386, 14/1/1917.

58 Spencer Tucker and T. J. Weiler, 'The Arctic Theater', in Spencer Tucker and Priscilla Mary Roberts, *World War I: The Definitive Encyclopedia and Document Collection* (Santa Barbara, Calif., 2015), 127–128.

59 NA, ADM 137/1386, White Sea, f/e 14 February 1917.

60 Ibid., f/e 12 December 1917.

61 Ibid., f/e 27 March 1917.

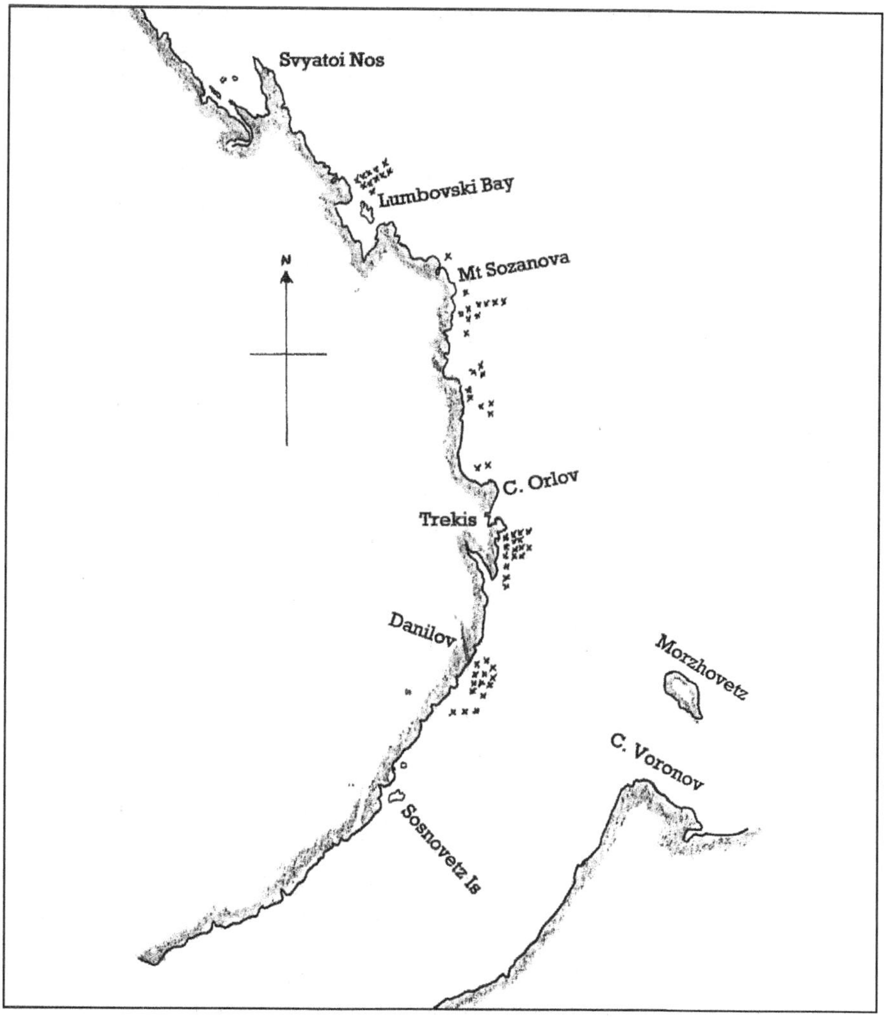

Figure 9: Outline map of the White Sea approaches
Source: Ken Knox

On 25 March 1917, another force of twelve trawlers had left Lowestoft and reached the Kola Inlet on 8 April.[62] Although their voyage north had been

[62] Some trawlers which had previously seen service in the Dardanelles and the rest of the eastern Mediterranean and had returned to Britain for a refit in October 1916 were apparently afterwards amongst the vessels sent to the White Sea. See Hurd, *History of the Great War*, 2:173.

uneventful, they were soon on high alert. The following day there were several reported sightings of a U-boat, and a Russian steamer exploded and sank off Syedlovati Island in the Kola Inlet. All the trawlers were soon despatched to the inlet for either sweeping or patrol duties. On 10 April, two mines were found in the vicinity of the recent sinking, and a few days later, when a group of trawlers led by the *Sir Mark Sykes* proceeded to investigate a sighting of another five or six mines, the *Arctic Prince* struck a mine which blew off its bows. Boats rescued the survivors, but six crewmen were killed. However, the trawler improbably remained afloat and was towed to safety by the *Sir Mark Sykes*, while the other trawlers destroyed one of the mines and dropped dan buoys to mark the location of the others which were cleared a week later.[63] German records subsequently revealed that 23 mines had been laid that year in the Kola Inlet, of which about 18 at most were located at the time. Another steamer, this time Norwegian, was lost before minesweeping operations ceased for the year.[64]

The flotilla command thought U-boats were using a secret base in the Norwegian fjords for their operations and on 17 April the *Sir Mark Sykes* led a force of five trawlers on a patrol in the vicinity of Varanger Fjord, supposedly taking care to stay outside territorial waters (although at this time British and Norwegian definitions of such limits differed). They stayed in the area for three days, and although no U-boat was found a vessel (which later proved to be the German steamer SS *Johanna Blumberg*) was spotted and chased, but escaped by slipping into Norwegian limits. The incident afterwards provoked considerable activity and controversy ashore, and the skipper of the armed trawler *Lanercost* refused to allow crew from a Norwegian patrol vessel, one of several craft which came out to investigate, permission to board.[65]

On 27 April, two further trawlers arrived in company with HMS *Intrepid* to reinforce the northern flotilla, followed shortly afterwards by another three, along with the Armed Yacht *Sagitta*. Meanwhile, a group of trawlers, under Gore Langton's command, set off from the Kola Inlet to patrol the Norwegian coast between Nordyn and Vardo in search of a U-boat that reportedly had been seen on several occasions. This patrol was maintained, amidst heavy winds and snowstorms, for six days, and though no submarine was spotted Gore Langton attempted to capture the SS *Vesta* of Hamburg on the 22nd, firing two shots across its bows, but the German vessel escaped by closing on the Norwegian coast and territorial waters.[66]

Throughout the remainder of the spring and into the summer, anti-submarine patrols and minesweeping operations were continued in the White Sea and adjacent waters, while ashore Russia was convulsed by political tumult. The

---

63 NA, ADM 137/1386, White Sea, f/e 22 April 1917.

64 NA, ADM 186/604, 'History of British Minesweeping in the War', 69.

65 NA, ADM 137/1386, White Sea, f/e 22 April 1917.

66 Ibid., various, f/e 6 May 1917.

March Revolution, which brought about the end of the Romanov dynasty and its replacement by a provisional government, pledged to continue the war seemed at first an outwardly positive move, but the region where the trawlers were based was, like the rest of Russia, descending into increasingly turbulent and unpredictable times. It was ever more important that the naval forces operating in these remote waters were as self-sufficient as possible. In August, a consignment of sheep was brought in from Britain to augment the supply of fresh food, as it was noted that pasturage was quite good in the region at that time of the year.[67]

In previous years, as we have noted, the main northern port of Archangel was only open to shipping between the months of July and October, but in 1917 the ice-free port of Murmansk was finally connected to the capital Petrograd by rail, and that year enormous quantities of military supplies and fuel were shipped in through both ports. The flotilla maintained its presence as the year moved towards its close, even though the second revolution in October had brought the Bolsheviks to power. On 19 November, 16 trawlers were based on Yukansie and kept busy escorting westbound shipping from the Archangel bar as far as Norwegian territorial waters, while five others were by then at Murmansk carrying out sweeping and patrolling duties in the mouth of the Kola Inlet. As the weather closed, the Archangel convoying ceased, and the size of the White Sea force ran down.[68] The crews returning home were granted three weeks' leave.[69] By 1 December 1917, only the *Sir James Reckitt* and six other trawlers were left on station. They faced another harsh winter and had to deal with the increasingly isolated and uncertain situation which followed the Bolsheviks taking power in Petrograd and Moscow.

What is clear is that, whether on active service in far northern waters, the Mediterranean, or elsewhere, Auxiliary Patrol vessels and their mainly fishermen crews found themselves serving during the first three and a half years of the Great War off a wide spread of European coasts, in waters far removed from the British Isles and on seas and oceans quite different from those in which most then plied their peacetime fishing activities.

67 Ibid., f/e 12 August 1917.

68 NA, ADM 137/1387, White Sea, various, f/e 4 December 1917.

69 Ibid., various, f/e 30 November 1917.

Chapter Six

Fishing during the Great War

Pre-War Position

Taken as a whole the British fishing industry boomed in the years preceding the outbreak of the Great War, and 1913, the last full year of peace, proved no exception. Most previous performance indicators in both the steam trawling and herring fisheries – the leading sectors in this dynamic trade – were surpassed. Total fish landings at British ports that year were well over the 16 million cwt (800,000 ton) mark, attracting revenue of more than £10 million.[1]

Whether judged by the size and sophistication of its fish processing, catching, or distributive sectors, the British fishing industry led the world. Flotillas of middle and distant water steam trawling fleets, working principally from the ports of Hull, Grimsby, Aberdeen, Fleetwood, Shields, and Milford Haven, took white fish from as far afield as fishing grounds off the Faroe Islands, Iceland, or the Barents and the Irish and North Seas as well as the English Channel and beyond. Most of their catch was destined for home markets and intended to satisfy a seemingly insatiable domestic demand, particularly from the burgeoning fried fish and chip trade. Meanwhile, large fleets of drifters voyaged up and down the coast intent on exploiting the immensity of herrings which shoaled in great quantities at different times in various seas off the British Isles. Much of the herring catch was processed by the highly mobile (and mainly Scottish) curing gangs which followed the drifters, and although some fish was smoked for domestic consumption as kippers, vast quantities were salt cured, packed in barrels, and exported to markets across immense areas of central and eastern Europe. The number of fishermen and boys regularly and occasionally employed in England and Wales in sea fisheries in 1913 were 37,870 and 7,512 respectively, while the total figures for Scotland were 38,262.[2] Those regularly or occasionally employed in the catching sector of the United Kingdom's fishing industry, including the Isle of Man and the Channel Islands, was just under

1 Robinson, *Trawling*, 141,

2 British Parliamentary Papers, Thirty-Second Annual Report of the Fishery Board for Scotland, Being for the Year 1913, xxxi, 16.

100,000 in 1913.[3] Many, many thousands more worked in the processing and distributive branches.

During the first half of 1914 there was every prospect that the prosperity of recent years would be maintained throughout the twelve months and, indeed, when Britain declared war on 4 August, many sectors of the trade were still working at full stretch. In the North Sea, large numbers of drifters were busy exploiting the summer herring fisheries off the coasts of Scotland and north-east England. Hundreds of steam trawlers were spread across fishing grounds around the British Isles and beyond. Those fishing in the North Sea were working either singly or in one of the four great Hull boxing fleets that worked far from land, but were serviced almost daily by fast steam cutters that then ran box-loads of newly caught fish straight for the Thames in order to catch the best prices on the next morning's market at London's Billingsgate. More than 200 old Lowestoft sailing smacks were at sea trawling on their traditional grounds between the Dowsing and the Gab. The seas around Britain were alive with fish and with the fishermen who made their living from them.

August 1914

Even though the European situation darkened and rapidly deteriorated as July ran its course, few people thought that the British would soon be at war. The suddenness of the declaration caught many people, organisations, and institutions unawares: not least the fishing industry, which had far more vessels at sea than safe in harbour when hostilities opened. Lamps might be going out all over continental Europe, but the lights of the fishing vessels still burned brightly across the North Sea on the evening of 4 August. At 8.15 p.m. that day, less than three hours before Britain declared war, Admiralty messages were telegraphed to the harbour masters of every east coast port with instructions that no fishing vessels should sail for North Sea fishing grounds, and that all such craft then at sea were to be directed by wireless to make port, if possible by dawn.[4]

Such abrupt orders were far easier to issue than to execute. Although some vessels, such as those in the Hellyer boxing fleet, received the news by radio; the use of wireless telegraphy, so far as the fish trade was concerned, was still very much in its infancy. For the most part, the communication had to be carried out to sea. At Lowestoft, for example, the instructions to return were taken to the port's sailing smacks working offshore by a steam trawler, but by the next morning there must have been numerous vessels still fishing far out in the

[3] British Parliamentary Papers, Fisheries England and Wales: Fisheries in the Great War: Being the Report on Sea Fisheries for the Years 1915, 1916, 1917 and 1918 of the Board of Agriculture and Fisheries, vii.

[4] Ibid., 1–2.

North Sea and completely unaware that war had broken out, while those craft working on distant water grounds, even when once informed, faced a voyage of several days before reaching the safety of their home port. One such craft was the Grimsby steam-line fishing vessel *Tubal Cain* which was taking fish off the west coast of Iceland when stopped by the German commerce raider *Kaiser Wilhelm der Grosse*, at 7 p.m. on 7 August. The skipper William Smith and the 14 crew only discovered that war had broken out when they were taken prisoner. The fishermen were allowed some time to collect their belongings and after the German commander asked if they had any fish on board Skipper Smith arranged for half a dozen fine halibuts to be supplied to their captors. Once on the German vessel the *Tubal Cain*'s crew were ushered below deck and not allowed to witness the demise of their vessel, then sunk by gunfire: probably the first British fishing vessel to be lost in the conflict. Many more would follow. The German ship then steamed south with its captives at 16 knots, and after sinking a couple of British merchant vessels was in its turn lost after being badly damaged in an action with HMS *Highflyer* on 26 August off the West Coast of Africa. By then the fishermen had been transferred to another vessel and were eventually landed at Las Palmas from where they made their way home. They can scarcely have imagined when they had left Grimsby for Iceland in late July that their trip would take them so far south.[5] Later in the war, Skipper Smith joined the Royal Naval Reserve (Trawler Section).

In this opening phase, some fishing vessels on both sides of the conflict had narrow escapes. The Grimsby steam trawler *Zenobia* was detained for several hours in Heligoland after being stopped by a German gunboat on 4 August, but then allowed to proceed. And just after midnight on 5 August three German trawlers and two drifters left Aberdeen shortly after the declaration of war came into force, just managing to get out to sea before a group of British officials rushed onto the quayside to arrest them. Not all vessels were so fortunate with either timing or communications: on the same day, the German trawler *Else Kunkel* and fishing smack *Hammil Warden* were interned after sailing into Aberdeen, unaware that war had been declared the previous night.[6]

By and large, all other British fishing vessels made it back to port, and there were no further losses to the enemy in the first two weeks of the war. The Admiralty had been keen to get fishing vessels off the North Sea. These waters were likely to be on the front line of the maritime war – no place, the naval authorities initially reasoned, for civilian fishing vessels. Indeed, it was also thought likely to be extremely difficult to protect unfettered fishing vessels, going about their daily business along the length and breadth of the British

---

5 *Grimsby Evening Telegraph*, 9 August 1939.

6 Fishing Boat Heritage: https://fishingboatheritage.co.uk/?option=com_content&view=article&id=9293:week-1%25E2%2580%25934th-to-8th-august-1914&catid=655&Itemid=174 (accessed 19 September 2018).

coasts and far beyond, and there was also a belief that they might furnish an avenue for smuggling enemy agents into the country, perhaps even saboteurs. Moreover, it was soon evident that substantial numbers of fishermen and fishing vessels would be required for wartime naval security duties. Indeed, those who already had military obligations were quickly mobilised. This was to have major effects on both the industry and coastal communities: on the Isle of Lewis, for example, 2,000 fishermen, a substantial proportion of the male population, were already in the Royal Naval Reserve.[7] Thus, the first intention of the Admiralty was to confine many civilian North Sea fishing vessels to port for the duration, but while it was soon seen as priority both to maintain effective security at sea and make the best military use of this reservoir of skilled seamen and ships, it was also quickly evident that, being an island nation heavily dependent on imports in peacetime, it was going to be necessary to maximise our domestic food production if the war proved to be of long duration.

And the immediate cessation of fishing certainly had had a dramatic impact. At that time, the bulk of the nation's fish was taken in the North Sea and supplies were quickly curtailed. Not only were fishermen and fishing vessels laid up, but thousands of ancillary workers were thrown out of work along the entire length of Britain's east coast. Large numbers of women employed by Scottish herring curers suddenly found themselves unemployed during the very part of the year when their opportunities to earn money were usually at their greatest. In big trawling ports such as Hull and Grimsby, where large numbers of vessels were normally always at sea, it was difficult to find berths for all vessels when they returned *en masse*. In Hull, the newly opened King George Dock and the old eighteenth-century Queen's Dock were both called upon to berth large numbers of trawlers and relieve the congestion in St Andrew's Dock, the normal base of the port's fishing industry.[8] After the initial shock was over there was soon a general wish to return to sea, but by then there was also the problem of obtaining insurance, as underwriters had withdrawn cover following the onset of hostilities.[9]

The Board of Agriculture and Fisheries and the Scottish Fishery Board soon made representations to the Admiralty, and there quickly followed a series of what were described as 'long and anxious conferences'.[10] Just over a week after the war had commenced, fishing vessels began returning to North Sea grounds (although for many of the larger vessels the issue of insurance still

[7] David T. Jones, 'Scottish Fisheries during the War', in David T. Jones, Joseph F. Duncan, H. M. Conacher, and W. R. Scott (eds), *Rural Scotland during the War* (London, 1926), 32.

[8] Robinson and Hart, *Viola*, 68.

[9] British Parliamentary Papers, Fisheries England and Wales: Fisheries in the Great War: Being the Report on Sea Fisheries for the Years 1915, 1916, 1917 and 1918 of the Board of Agriculture and Fisheries, xxi and 75.

[10] Ibid., 2.

had to be clarified and this delayed numerous sailings for a short while longer). The Admiralty remained concerned, and a long series of naval regulations were gradually formulated to keep the movements of fishing vessels under some sort of control and ostensibly to ensure they kept out of several sensitive or dangerous areas, including newly sown minefields.[11]

The fishing proved good for much of the remainder of the summer, and most of those involved benefited from buoyant white fish prices, but the costs of continuing fishing quickly became all too evident. Soon after the resumption of activities it was clear that the waters to which the fishermen had returned were now fraught with additional hazards. Although the *Tubal Cain* was the only civilian fishing vessel to have succumbed to an enemy attack during the first couple of weeks of war, all this changed from the night of the 21–22 August. That evening, eight trawlers were taken by the enemy vessels supporting German minelaying activities. By 26 August, 14 Grimsby and ten Boston steam trawlers had been sunk and their crews taken back to Germany as prisoners. Then, on 27 August, the drifter *Barley Rig* struck a mine off the Tyne with the loss of five crew, the remainder being picked up by the Hull steam trawler *St Clair*.[12] Another five fishing vessels succumbed to either the Humber or Tyne minefields during the month of September and at least 25 civilian fishermen lost their lives.[13]

Most of those fishermen taken prisoner in August 1914 were landed at Wilhelmshaven or other German ports and the treatment meted out to these non-combatants on arrival was sometimes horrific. One Boston fisherman described being jeered at and spat at by crowds as they were marched through a town after landing and a number succumbed to illness or ill-treatment during the early days of incarceration. There were several examples later reported of brutality to those held in such circumstances.[14] Eventually, they were confined with interned merchant seamen and others mainly in Ruhleben Camp near Berlin. Over 5,000 people were incarcerated in this civilian detention camp created on a former harness racing track, situated a few miles to the west of Berlin, although others went to Sennelager Camp. It was many weeks before relatives back home knew their fate. Throughout the war relief funds were organised in many fishing ports, and parcels of food, clothing, and cigarettes sent to those who were held. Many interned fishermen were repatriated at various times in the war, but others spent the duration of the conflict in captivity.[15]

---

11 Ibid., ix–x.

12 Hurd, *History of the Great War*, 1:322.

13 *British Vessels Lost at Sea, 1914–1918*, 99–100.

14 See, for example, *Hull Daily Mail*, 23 December 1915; *Sheffield Evening Telegraph*, 18 May, 1916; *Newcastle Journal*, 8 August 1916; *Globe*, 7 June 1916; *Boston Guardian*, 10 March 1917; *Boston Guardian*, 12 January 1918. Also 'Port of Boston, Lincolnshire: Where Fishermen Became Prisoners of War', BBC World War One at Home: https://www.bbc.co.uk/programmes/p022x1gw (accessed 9 September 2018).

15 See, for example, *Leeds Mercury*, 2 December 1918 and 8 January 1919.

The Last Months of 1914

Eighteen more civilian vessels were lost while fishing before the end of 1914, and, except for the motor fishing vessels *Constance*, *Wayside Flower*, and the steam trawler *Manx Queen* (sunk during the German raids on Hartlepool, Scarborough, and Whitby in December), all were destroyed by mines. The port which suffered the most casualties to mines in this period was Lowestoft, which lost four fishing vessels – *Fraternal*, *Will and Maggie*, *Speculator*, and *Lord Carnarvon* – and at least 24 men and boys died with them.[16] Several other vessels had narrow escapes during these months after reportedly fouling mines in their nets, but the war situation soon manifested itself to fishermen in other ways. When the U-9 sank the British cruisers *Aboukir*, *Cressey*, and *Hogue* in September 1914, many of the survivors were picked up by Tom Phillips and George Jacobs, who were respectively the skippers of the Lowestoft sailing trawlers *Coriander* and *JCG*. Disregarding any risk, both hurried their vessels to the scene as the great warships foundered.[17]

Similarly, skippers Ambrose Fisher and Frederick Woolaston of the Hull trawlers *Euripides* and *Cameo*, together with William Ilett, Henry Wicks, and Mark Howard, respectively skippers of the Grimsby trawlers *Silanion*, *Straton*, and *Prince Victor*, were presented with the silver award for gallantry by King George V after braving a minefield to rescue passengers and crew from the Wilson Line steamer *Runo*, mined off the Tyne, en route to Archangel on 5 September 1914.[18] They took their trawlers alongside the stricken vessel's starboard and port quarters and Ilett rescued 120 passengers, including all the women and children, while Fisher took a further 30 people on board, including Captain George Henry Lee, who was the last to leave the vessel. Many survivors were also picked up from the sea or lifeboats by the other trawlers, but 29 passengers and crew did not make it. Skipper Joseph Wood, who also participated in the rescue, lost his life when the Hull trawler *Imperialist* was mined off St Abbs Head the following day.[19] At the subsequent Board of Trade enquiry into the sinking, Captain Lee had his master's certificate suspended for six months for having altered course in direct contravention of government notices and owner's instructions, and this was said to have brought his vessel onto the minefield.[20]

---

16 I am grateful to Alan Curtis for these details.

17 Corbett, *Naval Operations*, 1:176.

18 *Manchester Courier and Lancashire General Advertiser*, 12 March 1915.

19 Thomas Wilson and Company also asked the Lord Mayor of Hull to award on their behalf gold watches to Ilett and Wollaston. See *Hull Daily Mail*, 24 September 1914.

20 British Parliamentary Papers, Board of Trade Wreck Report for 'Runo', 1914: https://plimsoll.southampton.gov.uk/SOTON_Documents/Plimsoll/21221.pdf (accessed 21 September 2018).

Fishing on the Maritime Front Line

Although most of the 26 fishing vessels lost in the first month of war had been captured by enemy surface craft – mainly during other naval operations – and their crews taken prisoner, the causes of the losses soon shifted. Nineteen of the 22 craft sunk while fishing between the beginning of September 1914 and the end of March 1915 were unwitting victims of mines, and at least 81 fishermen perished with them.[21] But, prior to the end of March 1915, there is little evidence that working fishing vessels were deliberately targeted in enemy attacks, and none seems to have been sunk by U-boats. All this changed again from the beginning of April. By this time the first German unrestricted warfare campaign was in full swing and the U-boats finally turned their attention towards civilian fishing vessels.

On 1 April 1915, the steam trawler *Jason* was working in waters some 40 or so miles north-east by east of the Tyne when a U-boat surfaced nearby and fired two shots across its bows. The trawlermen abandoned ship, scrambled into their small boat, and were taken on board the U-boat deck where they were held at gunpoint while the Germans rowed across to the trawler to lay charges. Once *Jason* was blown up, the U-boat then turned its attention towards the *Gloxinia* which had been fishing nearby. The *Gloxinia*'s crew were likewise ordered to board the U-boat, but the charges laid in the hull of this trawler proved faulty and even though a number of rounds were fired at the vessel it still refused to sink. The Germans boarded once more, laid further charges and finally finished off the trawler. Not content with two sinkings, the U-boat commander then bore down on the trawlers *Nellie* and *Rhodesia* that were fishing some five miles away. He went alongside the *Rhodesia*, transferred aboard the crews of both the *Jason* and *Gloxinia* and told the skipper to stand by to pick up the crew of the *Nellie* which he then chased, captured, and finally sank with gunfire. The U-boat commander returned once more to the *Rhodesia* and informed the skipper that the *Nellie*'s crew were in their rowing boat some way off his starboard bow. Despite searching for hours, the *Rhodesia* could not find the missing fishermen and eventually gave up and headed back to the Tyne. Later, the exhausted crew of the *Nellie* were picked up by another fishing vessel and brought safely to shore. The *Jason*'s crew reported being treated with extreme civility while held captive, being provided with bread, coffee, and tobacco.[22] A few days later, the same submarine reportedly sank yet another trawler, this time the *Ancantha*, some 25 miles east by north of the Longstone.[23]

The manner of these trawler sinkings provided something of a template for many subsequent U-boat attacks on working fishing vessels during the Great

21 *British Vessels Lost at Sea, 1914–1918*, 99–101.

22 *The Times*, 3 April 1915.

23 See U-10 uboat.net: https://uboat.net/wwi/boats/successes/u10.html (accessed 19 September 2018).

War. Although mines continued to send fishing vessels to the bottom throughout the rest of the conflict and though there were occasional further losses to surface raiders, by far the majority of such craft sunk between April 1915 and the Armistice succumbed after being captured by enemy submarines. According to one official source, in the months of May and June 1915 alone, at least 80 fishing vessels out of a total of 92 fishing craft lost through enemy action were captured and sunk by submarines and in most of cases the crews survived, being usually allowed to sail for shore or to other vessels in their open boats.[24]

At this juncture, Hull's Hellyer Steam Trawling Company was probably hit harder than any other large fishing firm, and by the end of July 1915 it had lost some 15 unarmed trawlers as well as a further two craft sunk on Admiralty service: indeed, almost one-third of the company's fleet was destroyed in that first full year of the war. At least one Hellyer skipper, Herbert Johnson, was unfortunate enough to be sunk twice by U-boats in 1915. The first occasion was on 15 May when his trawler the *Hector* was sunk in the North Sea. Two and a half months later, by then skipper of the *Cassio*, Johnson and his crew were cast adrift once more in an open boat after his new trawler was chased, stopped, and sunk off the Shetland Islands while returning from an Icelandic trip. On the first occasion, the U-boat commander had ensured that the trawler's provisions were placed with the fishermen in their open boat and had given the skipper some bread as well, and survivors of a number of such sinkings sometimes remarked on other instances of civility. When the Aberdeen-owned trawler *Lucerne* was stopped and sunk by a U-boat some 50 miles north by east off Rattray Head, for example, the German officers made two trawlermen row them over to the trawler where they not only laid bombs but also helped themselves to newly caught fish. Afterwards, they put the Aberdonian trawlermen aboard a Danish vessel but not before the Germans insisted on paying two marks for the fish they had taken.[25]

There were a number of mass sinkings of fishing vessels around this time, as, for example, on the night of the 23–24 June 1915, when the submarine U-38 captured and scuttled 16 fishing vessels, mainly drifters, most of which had been working about 40 or so miles to the eastwards of the Outer Skerries off Orkney.[26] On the night of 30 July another group of fishing vessels, this time nine Lowestoft trawling smacks, were sunk by the UB-10 from the Flanders Flotilla while working some 20 or so miles from the East Anglian coast. One of those lost was the *Coriander*, which had played a part in the rescue of survivors from the sinking of the *Aboukir*, *Cressey*, and *Hogue*. On both occasions, in what was

---

24 *British Vessels Lost at Sea, 1914–1918*, 102–105.

25 George F. Ritchie, *The Real Price of Fish: Aberdeen Steam Trawler Losses, 1887–1961* (Beverley, 1991), 20.

26 The U-38 also sunk the Norwegian steamer *Truma* on the same day. See uboat.net: https://uboat.net/boats/successes/u38.html (accessed 19 September 2018).

already usual practice after such attacks, all the fishermen were allowed to row away in their open boats and survived.[27]

Fish was clearly being perceived as a strategic resource, and the Germans were by now not only trying drastically to curtail Britain's food supply, they were also attempting to eliminate the remaining civilian trawlers and drifters that might also be taken into war service by the Admiralty. There was now clearly no doubt that by this time non-combatant fishermen and their unarmed trawlers were carrying out their daily business on what had become the maritime military front line.

Because they were working in what were to all intents and purposes maritime war zones, civilian fishermen were often involved in rescuing survivors of vessels sunk in the conflict. In May 1915, for example, Skipper William Ball and his six-man crew of the Isle of Man fishing vessel *Wanderer*, which had been working off the south coast of Ireland, witnessed the torpedoing of the liner *Lusitania*. They found people screaming and fighting for their lives in the water and managed to pull 160 desperate survivors on board their small craft and took two full lifeboats under tow.[28] Fishermen also picked up countless numbers of colleagues whose fishing vessels had been sunk by the German submarines (Photograph 8).

The carnage wrought by U-boats amongst the fishing fleet continued into late September 1915 after which this first wave of attacks subsided and only one further civilian fishing vessel, the trawler *King William*, was lost during the remainder of the year – and that fell victim to a mine. Even so, 192 civilian fishing vessels, mainly trawlers and drifters, had been captured or sunk during the whole year. These losses had been geographically concentrated: all but a handful being sunk in the North Sea, or near the Orkney and Shetland Islands. In January 1916, the U-boats resumed their attacks on the fishing fleets and a further 132 were sunk during the ensuing year.[29] Virtually all were destroyed by submarines and by this time U-boat commanders seemed to have become yet more adept at engineering wholesale sinkings, sometimes managing to destroy large numbers of fishing vessels in one attack. Late on the evening of 24 September 1916, for example, the U-57, commanded by Carl-Siegfried Ritter, surfaced close to the Scarborough trawler *Fisher Prince* which was hauling its catch on the Whitby Fine Grounds. Using grappling hooks, the Germans quickly boarded and took control of the vessel, locking up the crew in the fish room. Ritter then overwhelmed the trawler *Otterhound* in a similar fashion. Nearby vessels were then picked off, one at a time. Quietly and quite covertly, for fear of scattering its prey, the German vessel closed in and then captured

---

27 *British Vessels Lost at Sea, 1914–1918*, 106; and U-10 uboat.net: https://uboat.net/wwi/boats/successes/ub10.html (accessed 19 September 2018).

28 David Ramsey, *The Lusitania Saga and Myth* (Barnsley, 2015), 72.

29 *British Vessels Lost at Sea, 1914–1918*, 101–113.

Photograph 8: Scarborough trawlermen held on the deck of a U-boat
Source: George Westwood

a total of 13 Scarborough trawlers, each time imprisoning their crews on the *Fisher Prince*. Only once all the trawlers were taken did the sinkings begin: the *Nil Desperandum* was sunk with a scuttling charge while the others were despatched by gunfire. By then Ritter had 126 captive fishermen on the *Fisher Prince* but, despite the not inconsiderable danger to himself and his crew, he stayed by the trawler all night and then stopped a Norwegian steamer soon after dawn. He parted company after transferring his captives aboard but not before finally sinking the *Fisher Prince*. Afterwards, the Norwegian vessel took the trawlermen into South Shields.[30] Ritter was to survive the Great War, sinking a total of 76 ships before the Armistice. The U-57 also outlasted the conflict only to be broken up at Cherbourg after being surrendered to the French.[31]

Not all similar encounters with unarmed fishing vessels turned out to the advantage of the U-boats. On 16 January 1916, not long after the resumption of German attacks, the smack *Acacia* of Lowestoft was trawling in company with other craft when a U-boat surfaced and opened fire with a machine gun. James Crooks, skipper of the *Acacia*, had been sunk by the Germans the year before and was determined not to suffer a similar fate again. He waited for a suitable moment as the U-boat closed in and then cut the trawl warp, put his helm hard over, and tried to ram his attacker, only missing its hull by a few feet. The action initially disconcerted the Germans and the *Acacia*, taking advantage of a fine breeze, ran for freedom and for the port of Lowestoft.[32] Crooks was later awarded the DSM for his actions.[33]

Enemy Attacks on Fishing Vessels beyond the North Sea

Most fishermen who fell victim to U-boat sinkings were afterwards allowed their liberty, but skipper James Strong and the crew of the Lowestoft smack *Veronica* found themselves being held on a U-boat for a few days. Their craft was captured and then sunk by the UC-16 off Start Point in Devon on 11 November 1916, and they spent around three days below decks. Later, they reported being well treated and able to fraternise with the Germans who provided them with tobacco when they were finally allowed to leave, the only degree of personal hostility they encountered on board being when the Germans brought up issues relating to the Royal Navy Q ship HMS *Baralong* and an incident in which the skipper was reported to have executed stranded German sailors. Their short stay on board was certainly eventful: on the first day of their detention, the U-boat encountered an English patrol vessel in the fog and, after making an emergency dive in shoal water, struck the seabed and tilted over at an angle of 45 degrees. The following day, the U-boat captured and then scuttled a French schooner before firing on a passing steamer after which it had to make another emergency dive when the merchant ship returned fire. The next day, another French schooner was captured and sunk, but on this occasion the fishermen were allowed to join the schooner's crew in their open boats. After five hours at sea, they were picked up by a French fishing boat and taken into Île de Sein,

30 Robinson and Hart, *Viola*, 99–100.

31 See U-57 uboat.net: https://uboat.net/wwi/boats/successes/u57.html (accessed 19 September 2018).

32 Robinson, *Trawling*, 135.

33 *Pall Mall Gazette*, 30 December 1919.

off Finisterre.[34] The UC-16 was lost around 11 months later, believed mined off Zeebrugge in October 1917.[35]

On 28 November 1916, two U-boats were involved in attacking fishing vessels in the English Channel off the south coast of England. The UC-21 Commander Reinhold Saltzwedel sank the trawling smacks *Clematis* and *Vulcan*, from Ramsgate, which were fishing between about 28 and 35 miles off the coast of Devon,[36] while the Commander of the UB-37 captured and then scuttled four Brixham sailing smacks, *Amphitrite*, *Catena*, *Provident*, and *Sea Lark*. The smacks *Diligence* and *Lynx* were also damaged by UB-37 but survived the initial attack and made it back into Brixham, although the former was later broken up as a constructive loss. Another smack, the *Terminist*, managed to escape by cutting its trawling gear and running clear when the U-boat appeared.[37] The UB-37 sank a further smack, the *Concord*, a couple of days later around 28 miles SE of Start Point. In all cases, the fishermen made it back to shore, but their attackers did not see out the following year. Gunther died when the UB-37 and all hands were sunk by the Q ship *Penshurst* in the Channel on 14 January 1917. The UC-21 disappeared somewhere in the Bay of Biscay in September 1917,[38] while its former commander Saltzwedel lost his life after running into a mine on the UB-81 off the Isle of Wight on 2 December 1917.[39]

Although North Sea fishing vessels remained the primary focus of German attention, such attacks on the fleets working in the south-west were a precursor of things to come. Indeed, the number of U-boat attacks off other British coasts increased in 1917 after the Germans resumed their strategy of unrestricted submarine warfare. On 30 January, the U-55 sank four Lowestoft trawling smacks and another from Milford Haven as well as the Belgian steam trawler *Marcel.* These vessels were fishing some 20 to 30 miles north-west of Trevose Head in Cornwall, and on this occasion the skipper, James Rodmell, and cook, Herbert Herring, of the Lowestoft trawler *Trevone*, were drowned. The trawler's crew had been ordered to row alongside and board the U-boat but both men

---

34 NA, ADM 137/3970, *Home Water Attacks, September 1916–September 1916.* I am grateful to Alan Curtis for supplying details of this encounter.

35 See UC-16 uboat.net: https://uboat.net/wwi/boats/successes/uc16.html (accessed 19 September 2018).

36 The Auxiliary Patrol drifter *Pelagia* was also sunk by a mine laid by the UC-21 off Nab Head on 28 November 1916 with the loss of 12 Trawler Reserve crew. See The Wreck Site, HMD *Pelagia*: https://www.wrecksite.eu/wreck.aspx?160210 (accessed 19 September 2018).

37 Brixham in the Great War: http://brixhammuseum.uk/onewebmedia/WWI_Booklet_LowRes_A4.pdf (accessed 19 September 2018).

38 UC-21 uboat.net: https://uboat.net/wwi/boats/?boat=UC+21 (accessed 19 September 2018).

39 Reinhold Saltzwedel, uboat.net: https://uboat.net/wwi/men/commanders/283.html (accessed 19 September 2018).

were washed into the sea. James Rose, the trawler's mate, later claimed that they could easily have been rescued but said the Germans took no notice, even though the skipper was seen hanging onto the submarine's aft side for over ten minutes before he was lost from view. The three surviving fishermen were taken below decks, fed, and given hammocks, then kept on board until the following day, when they were put into the open boats of a French sailing ship which had just been captured.[40]

After the war the U-boat's commander, Wilhelm Werner, was accused of a number of war crimes involving various sinkings but never stood trial.[41] A couple of days after destroying these five fishing vessels Werner also sank the Ramsgate smack *Ada* and the *Inverlyon* of Lowestoft. The *Inverlyon*, having previously worked as a Q ship, had sunk UB-4 in August 1915.[42]

Another twelve sailing smacks were sunk off the south-west coasts on 11 and 12 March 1917. Ten of these vessels were captured and sunk by Paul Hundius, commander of the UC-47 off Trevose Head, while two others, the *Forget Me Not* and *Reindeer*, were scuttled after being taken by the UC-66, commanded by Herbert Pustkuchen. The majority of these vessels were owned and registered in Lowestoft or Ramsgate and, ironically, were probably working in the south-west to avoid the carnage being wrought by U-boats in the North Sea. The fishermen also made it ashore but neither of these U-boats was to survive the war nor their then commanders.[43]

There was yet another mass sinking of sailing trawlers off the south-west coast less than two weeks later. This time another nine, again mainly Ramsgate- and Lowestoft-owned vessels, were captured and sunk by the UC-17 commanded by Ralph Wenninger when working between 12 and 14 miles south-east of the Eddystone. Although further sailing trawlers, sometimes small groups of craft, were sunk with monotonous regularity in the south-west throughout the rest of the year, there were no more wholesale sinkings in the region on quite this sort of scale.[44]

---

40 ADM 137/3977, Enemy submarines, attacks on merchant vessels in home waters. I am grateful to Alan Curtis for access to this material.

41 TheWreck Site, FV *Trevone*: https://www.wrecksite.eu/wreck.aspx?147033 (accessed 19 September 2018).

42 U-55 uboat.net: https://uboat.net/wwi/boats/index.html?boat=55 (accessed 19 September 2018).

43 U-boat Commander Listing uboat.net: https://uboat.net/wwi/men/commanders/listing.html (accessed 19 September 2018).

44 *British Vessels Lost at Sea, 1914–1918*, 118–121.

Fishermen and Fishing Operations during Final Period of Unrestricted Submarine Warfare

Faced by such levels of attrition, in early 1917 the Admiralty assented to requests from Lowestoft smack fishermen that some of their vessels be armed. A couple at least became what were basically Q ships, and several epic actions with U-boats then took place which belied the technological gulf between such craft.[45]

At the beginning of February 1917, on the day the Germans opened their final unrestricted submarine warfare campaign, two Lowestoft smacks, the *I'll Try* and *Boy Alfred*, one powered by an auxiliary motor and now carrying both arms and naval ratings, were hauling their trawls when they sighted two distant submarines heading towards them. The smacks continued hauling but readied their armaments. As the first U-boat closed in, a figure in the conning tower opened fire with a rifle and hailed the *Boy Alfred*'s crew, calling for them to abandon ship. By this time the enemy was only 150 yards away and the skipper, James Wharton, quickly ordered the gun unshipped and firing commenced. The *Boy Alfred*'s third shot hit the submarine, which swiftly submerged. Meanwhile the other submarine which had turned on the *I'll Try*, was also reportedly hit by a shot from its would-be prey and likewise submerged with all speed.[46]

That summer, Tom Crisp, skipper of the *I'll Try*, which had by now been renamed *Nelson*, engaged in another fight with a submarine, but on this occasion the Germans got the better of the action. Crisp and his crew on the *Nelson* used their 13-pounder gun to put up a valiant fight, but were out-ranged by the German's superior armament. The fourth shot from the submarine went through the port bow of the fishing smack just below the waterline but the exchange continued. The seventh German shell hit Tom Crisp, passed through the side of his body, then smashed through the deck before exiting the vessel. The skipper was mortally wounded but continued giving orders until his final minutes and then his son took charge, eventually taking the crew to safety in the open boat as the stricken smack sank beneath the waves. The U-boat then turned on the accompanying armed smack, the former *Boy Alfred* now renamed the *Ethel and Millie*, which was first captured and then sunk.[47] The *Nelson*'s crew spent two days at sea in their open boat until they were picked up after mooring to a buoy some 40 or so miles from Lowestoft. Tom Crisp received the VC for his actions and this is believed to be the only time that father and son have been present at an action which led to the award.[48] The crew of the *Ethel and Millie* were last glimpsed when lined up on the U-boat deck but never seen again.[49]

---

45 Robinson, *Trawling*, 139–140.

46 NA, ADM 137/3318, Loss of Special Service Smacks, Ethel and Millie, Court of Enquiry.

47 Ibid.

48 Robinson, *Trawling*, 139–140.

49 Gibson and Prendergast, *German Submarine War*, 199.

In late spring 1917, a more comprehensive response to German attacks on fishing craft, known as the Group System, was introduced. All offshore fishing vessels were hitherto placed under naval control and ordered to fish together in groups. A few trawlers in each group were armed, and the intention was to fit at least one vessel with wireless as soon as these became available. The craft were commissioned and flew the White Ensign while the crews were eventually placed in a newly created Special Fishery Reserve. For the purposes of this protective scheme all such fishing vessels were regarded as naval units and therefore came under the commands and orders of the SNO responsible for the ports they operated from. It was as if they were now considered to be units of the Auxiliary Patrol and, although SNO took advice from key players in the local fish trade, he could not only decide the grounds on which the fishermen worked but also, in theory, their times of arrival and departure from port.[50]

Such arrangements were not confined to fishing vessels working the North Sea or voyaging to Iceland and Faroe. In south-west England, a somewhat similar system of protection was introduced for sailing trawlers, which were accompanied to sea by armed auxiliary powered smacks. At Brixham, for example, the two largest fleets were afforded protection by a couple of armed smacks while a smaller fleet had its operations restricted to inshore waters. Plymouth-based vessels also fished under the protection of two armed smacks.[51]

Not surprisingly, many problems were encountered in the first few months after the scheme came into operation. Discipline in such fishing sections was, perhaps inevitably, a problem at first, particularly amongst the steam trawlers voyaging to Faroe and Iceland. It was soon reported from Grimsby that a couple of vessels, the *Sisapon* and *Olearia*, had been sunk off Faroe by a submarine after they had voyaged without protection despite what were described as 'plainly worded warnings'.[52] Even amongst vessels voyaging in protected groups there were issues and, after complaints from the Admiralty, the Grimsby skipper of one of the newly armed trawlers was suspended *sine die* for leaving his section, and a Hull skipper was also suspended, in his case for six months, for a similar offence.[53] The first armaments supplied to the fishing trawlers protecting each group were usually too light to offer decent protection, and in July 1917 a group of eight distant water Grimsby fishing vessels were captured and destroyed by U-boats while voyaging between the Pentland Firth and Faroe.[54] This led to a temporary suspension of these distant water fishing trips until some larger 12-pounder guns could be fitted, and voyages were resumed in early September

[50] British Parliamentary Papers, Fisheries England and Wales: Fisheries in the Great War: Being the Report on Sea Fisheries for the Years 1915, 1916, 1917 and 1918 of the Board of Agriculture and Fisheries, 15.

[51] Ibid., 17.

[52] NA, ADM 137/719, Auxiliary Patrol Humber, 26 May 1917.

[53] Ibid., 19 May 1917.

[54] Ibid., 14 July, 1917.

1917. By November 1917, these sections consisted of six trawlers, five of which were armed, two with 12-pounders and one with wireless.

But, overall, as initial teething problems were ironed out and the vessels equipped with better armaments, the new arrangements proved their worth. Fishing craft sinkings were drastically reduced. During the first six months of 1917, 185 British fishing vessels were sunk by German U-boats, with 43 being sunk in the month of March alone. In the second half of the year, only 35 such craft succumbed in a similar fashion. The new armed arrangements were reported to be particularly effective amongst Humber-based trawlers working in the North Sea. In November 1917, the SNO Humber reported that since the system had been set up, trawlers from the ports working in the North Sea sections had completed 4,000 fishing voyages, covering a total of 1,200,000 miles while working on the fishing grounds or steaming to and from port. During this period, the sections had engaged in actions with the enemy on 17 separate occasions with the loss of just one trawler. These North Sea groups were also supported on occasions by a couple of the more heavily armed orthodox Auxiliary Patrol trawlers. A total of 288 voyages had also been made to Iceland with a further 32 to Faroe by the Humber trawlers.[55]

Certainly, the effectiveness of the new Fishery Reserve arrangements in terms of the distant water sections or groups was demonstrated in the early summer of 1918. Shortly after 5 a.m. on 20 June, a section of six trawlers, five of them from Hull, in voyaging home from an Icelandic trip, were confronted by a large German U-boat, U-53, around 55 or so miles to the southwards of the Faroe Islands. As the trawlers responded by taking up a line ahead formation, the German craft opened fire on the leading vessel, the armed trawler *Conan Doyle*. Aboard the *Conan Doyle*, the section's commander, Lieutenant John Francis MacCabe, RNVR, returned fire, and, after about half an hour of exchanges, scored a direct hit on the assailant. The U-boat briefly broke off the attack but soon returned to the fray, firing again, then lost contact once more when a heavy rainstorm swept in, enveloping both sides. This proved but a short respite: the weather quickly cleared, and the U-boat and *Conan Doyle* once again resumed their deadly exchanges, while the rest of the trawler group swung round to the south to make best use of the wind. By now the U-boat was also pouring fire from its three guns into some of the other trawlers in the section, scoring several direct hits on the *Aisne*, killing one deckhand and wounding most of the others. The section's ammunition ran low and the situation looked increasingly desperate until a shot from the by now badly damaged *Aisne* appeared to strike the German's aft and the *Conan Doyle*, while readying to try a ramming, reportedly hit their assailant with two more shells. Whatever happened, the U-boat disappeared in a shroud of cloud and smoke. The attack was over, and the trawler section resumed their homeward voyage. In his report on the incident

[55] NA, ADM 137/720, Auxiliary Patrol Humber, 10 November 1917.

MacCabe attributed the principal credit for saving the convoy to Skipper William Addy of the *Conan Doyle* and was unequivocal in his praise for the trawler crews: 'This was a fishermen's fight, and without doubt they put up a right stout one. I deem it an honour to have been in command of them'.[56]

The U-boat may not have been as badly damaged as was thought at the time but it is clear that fishermen, under the command of one RNVR officer, had driven off their attacker.[57] The U-53 is now considered to have been one of the five most successful German Great War submarines in terms of the number of ships it sunk.[58] Afterwards, MacCabe was made a Companion of the DSO; five trawler skippers, William Addy, Joseph Parrish, Robert Syrett, John Moston, and Sidney Hewitt were awarded DSCs; while another nine trawlermen, including Sidney Beck of the *Aisne*, who died of his wounds, were awarded DSMs. A further 16 trawler crew members from the section were also mentioned in the *London Gazette*. Many of the awards were presented by Sir Edward Charlton, Vice Admiral Commanding East Coast of England, at a large gathering in the Council Chamber in Hull's Guildhall in February 1919. The Admiralty also awarded £500 in prize money, MacCabe receiving £30 17*s*. 0*d*., the skippers £18 10*s*. 0*d*. each, and the other members of the crews varying sums down to £3 14*s*. 0*d*.[59] Sir Arthur Conan Doyle later sent William Addy, skipper of the *Conan Doyle*, an inscribed silver cigarette case.

Just over a month later the *Aisne*, by now repaired, was involved in another action with a number of other armed trawlers, on this occasion when part of a fishing section working in the North Sea out of the Humber. This time an attacking U-boat was driven off after being engaged by several armed trawlers after a 20-minute encounter.[60] No vessels in the fishing section were reported to have been hit.[61] Other attacks were reported during the month but by now groups of working trawlers were proving to be anything but easy targets for the German submarines.

---

56 *Yorkshire Evening Post*, 12 February 1919.

57 British Parliamentary Papers, Fisheries England and Wales: Fisheries in the Great War: Being the Report on Sea Fisheries for the Years 1915, 1916, 1917 and 1918 of the Board of Agriculture and Fisheries, 53–54.

58 'The U-boats of World War One, 1914–1918', uboat.net: https://uboat.net/wwi/boats/ (accessed 19 September 2018).

59 *Hull Daily Mail*, 13 February 1919.

60 British Parliamentary Papers, Fisheries England and Wales: Fisheries in the Great War: Being the Report on Sea Fisheries for the Years 1915, 1916, 1917 and 1918 of the Board of Agriculture and Fisheries, 54–55.

61 NA, ADM 137/722, Auxiliary Patrol Humber, 26 July 1918.

Inshore Fishing Operations

Because so many of the steam trawlers and drifters were taken up for Admiralty service or else lost while fishing, the inshore sectors assumed a greater level of importance in terms of food supply during the conflict. Prior to 1914, as noted above, the inshore fisheries in many areas were apparently locked into what was a seemingly unrelenting cycle of decline. During the last decade or so of the nineteenth century, several district sea fisheries committees had been set up around the English and Welsh coasts to oversee activities in inshore waters and, where necessary, pass by-laws on such subjects as close seasons and types of fishing gear that could be used with the overall aim of trying to conserve the viability of fishing stocks in waters under their jurisdiction. During the war, attempts were made to improve the efficiency of the inshore fisheries with the primary aim of increasing the amounts of fish they contributed to the national food supply for the duration of the conflict.[62]

There were several short- and longer-term options available for temporarily improving the yield from British inshore fisheries. One was to suspend or override some of the legislation aimed at the longer-term conservation of stocks, and this happened in some districts. Inshore trawling had previously been banned in many coastal waters, but such restrictions were sometimes relaxed during the Great War. During 1915 and 1916, for example, the North-Eastern District Sea Fisheries Committee which superintended coastal waters from the Tees down to Donna Nook in Lincolnshire gradually eased restrictions on trawling for the duration in response to petitions from various groups of local inshore fishermen.[63]

But another major means of improving the efficiency of the inshore fisheries was motorisation. Prior to the war, few inshore vessels were motorised. This, in part, was due to a chronic lack of capital required to purchase and install the requisite engines, and partly to the dearth of mechanical skills amongst the local workforce. Even so, some developments had taken place. One notable pioneer in this field in south-west England was Stephen Reynolds (1881–1919), a fishermen, author, and spokesman for the inshore industry who worked out of Sidmouth in Devon. In 1909, he had designed and had built a small motor vessel, *Puffin*, but ironically this was taken into Admiralty service soon after the outbreak of war. In 1915, a couple of months after Brixham fishermen had helped in the rescue of sailors following the sinking of HMS *Formidable*, Reynolds brought to their port news that the Government would provide £2,000 for a project to fit auxiliary engine power to some classes of sailing trawler and held meetings with interested parties.[64]

---

62 Robinson, *Trawling*, 142.

63 Humberside County Record Office, North Eastern District Sea Fisheries Committee, Minutes, 28 July 1915, 10 November 1915, 9 February 1916, and 11 May 1916.

64 Samantha Little, 'The Brixham Fishing Fleet in the Great War', in H. French, *Food,*

Reynolds was clearly an influential figure in fisheries circles. He had served on a committee for inshore fisheries prior to the war which had recommended the installation of inboard engines in small sailing craft and the process probably got started in the south-west, thanks to a loan from the Development Commissioners. This initiative was taken up nationally by the Fishery Board after it received a grant of £50,000 from the Development Commissioners and a Motor Loan Committee was created, under the Chairmanship of Cecil Harmsworth. By the end of 1918, 152 loans had been issued to the amount of £28,350.[65] Even so, before this project got into full swing, the process of motorisation was already under way in several other ports. Motorisation of some smaller vessels seems to have started along the Yorkshire coast on the cusp of the outbreak of war. For example, by mid-November 1914, at least six herring mules working out of Bridlington had been motorised,[66] and by the end of 1915 it was noted that a considerable number of motor boats were working on this stretch of the north-east coast,[67] and numbers increased here throughout the rest of the war.

Many obstacles had to be overcome to ensure the effective diffusion of engines in the inshore fisheries. First, there was the need to ensure the most suitable engines were acquired and find or train suitably skilled personnel. Then there was the need to adapt and modify many different designs of inshore vessel. It was also necessary to counter the inbuilt conservatism prevalent in many inshore fishing communities and overcome an enduring aversion to debt in such villages. Another problem, which increased as more vessels were fitted with engines, was securing sufficient supplies of fuel: the Scottish Fishery Board attempted to overcome such problems by preparing advance estimates of quantities likely to be required by fishing fleets working north of the Border. These were forwarded to the wartime Petrol Control Committee which oversaw and prioritised the distribution of supplies. This was an initiative which seems to have met with some success.[68]

But the biggest incentive that drove through motorisation, once a small number of boats had been fitted with engines in any fishing station, was the greater returns that were made. Motorised vessels were much more efficient; they could reach the grounds more quickly than their sailing counterparts, voyage further afield, and haul gear more efficiently. At one south-west port, for example, it was noted in the Fishery Board's post-war report that a motor fishing vessel had towed its sailing companion out to the grounds and while the motor boat

Farming and Fishing in Devon during the First World War (Exeter, 2017), 83.

65 British Parliamentary Papers, Fisheries England and Wales: Fisheries in the Great War: Being the Report on Sea Fisheries for the Years 1915, 1916, 1917 and 1918 of the Board of Agriculture and Fisheries, 63.

66 Humberside County Record Office, North Eastern District Sea Fisheries Committee, Minutes, 11 November 1914.

67 Ibid., 31 December 1915.

68 Jones, 'Scottish Fisheries during the War', 63.

had brought back fish which sold for £310, the sailing vessel's catch realised only £50. These were evidently larger vessels engaged in deeper water fisheries, but the first two inshore motor vessels fishing out of Poole also illustrated this gulf in profitability when they were reported to have made more than £10 each from a day's landings while none of the port's sailing craft managed to earn over ten shillings for their catches.[69] Reports of such gulfs in profitability were noted in many parts of the country. By the end of the Great War the use of engines in the inshore fisheries was well established and the long-term displacement of sail and oar from almost every commercial branch of the sector was well under way.

Conflicting Resource Demands

One problem afflicting the inshore fisheries as much as the offshore sectors was the diminution in the labour force thanks to calls from the military. At the outbreak of war, many fishermen and ancillary workers, particularly those thrown out of work by the early disruption to North Sea fishing, responded to calls to join the army and were lost to both fishing and Navy. Others, like a substantial number in the Scottish Isles and the south-west, were already in the traditional RNR and were soon called up. Yet others, of course, particularly those working in the major fishing ports, were in the new Royal Naval Reserve (Trawler Section) or else joined up when their vessels were requisitioned by the Admiralty. Personnel problems in the fishing industry, particularly in the inshore sector, became especially acute after the introduction of conscription in January 1916, which applied to all single and medically fit men between the ages of 18 and 41. Although fishing was classified as a certified essential occupation, it was still necessary to attend a Military Service Tribunal to secure exemption, and if this found against them they could be taken for either the Navy or the Army. By all accounts the tribunal system was a somewhat inconsistent process, and in England the whole issue of conscription of fishermen created further tensions between the Fishery Board and the military authorities.[70]

Apart from their crucial role in maintaining the national food supply, fishermen clearly had seafaring skills that were potentially of more use to the navy than the army. If not immediately required for naval service and thus taken by the army at a tribunal then the navy would thus be deprived of accessing more men from a shrinking reserve that possessed skills and special aptitudes that had already proved most valuable in the war at sea. Fishery authorities on both sides

69 British Parliamentary Papers, Fisheries England and Wales: Fisheries in the Great War: Being the Report on Sea Fisheries for the Years 1915, 1916, 1917 and 1918 of the Board of Agriculture and Fisheries, 60.

70 N. Hyman, 'Inshore Fisheries in Wartime Devon', in French, *Food, Farming and Fishing in Devon*, 77–78.

of the border thus pushed for authoritative recognition of the principle that no fisherman should be taken by the army who was willing to serve in the Auxiliary Patrol, if and when called to do so. The risk of losing such a diminishing pool of seafarers to the Army as a result of conscription was also recognised by the naval recruiting authorities who were thus inclined to enrol large numbers of eligible fishermen in anticipation of future requirements and then train and retain them in barracks until actually required. This, of itself, was perceived by the fish trade as a wasteful use of resources and after further representations to the Admiralty from the Fishery Boards a solution was arrived at.[71]

This involved the creation of what was named Section Y of the Royal Naval Volunteer Reserve. Fishermen and other professional seamen who joined this section could continue their occupation for the time being, provided they accepted liability for service when called up. Section Y procedures reduced the likelihood of skilled seafarers being lost to army service and usually allowed them to continue their occupation until required by the Royal Navy. It was also overseen by the Fishery Boards in England and Scotland. Every fisherman enrolled in Section Y could appeal for a delay on being called up, and local fishery officers could report on the issues relating to individual cases. The main objects were to reserve key men for whom no substitutes were available, to ensure the most efficient boats left fishing had sufficient crews to continue their work, and to avert the extinction of any fishing community by preventing the conscripting of all its remaining fit fishermen.[72] The Boards were thus able to take a reasonably holistic view of the industry, regulating to a considerable degree the procedure of calling up men from this diminishing pool of labour, and, in so doing, ensuring the least detrimental effect on fish supplies. Although there were some complaints (for example, amongst Devonian fishermen about what they regarded as 'inappropriate conscription when sent for training at Devonport barracks when they could have been fishing'),[73] the system, which was later described as 'the pocket from which the navy drew men for the trawler section when required',[74] generally worked well.

Despite all the exigencies of war, the British fishing industry, albeit with a drastically reduced pool of labour, managed to continue working throughout the conflict. By the end of the war, fewer than 14,000 men were left fishing, of which over 8,000 were over military age, and 400 below. An increasing number of vessels were manned by a mixture of boys getting their first taste of life at sea, sometimes at an earlier age than usual, and by others continuing to fish well past retirement age. The official government report on wartime fisheries activities remarked on at least one trawler going to sea with everyone in the crew being

[71] Jones, 'Scottish Fisheries during the War', 38–39.

[72] Ibid.

[73] Hyman, 'Inshore Fisheries in Wartime Devon', 78–79.

[74] Jones, 'Scottish Fisheries during the War', 38.

over the age of 70. Although most reports tend to talk about men, some women and girls were also working in the catching sector.

The case of Ella Trout provides a rare glimpse into the activities of some women and children in the inshore fisheries during the Great War. Twenty-year-old Ella lived in the remnants of the Devonian village of Hallsands, which had been largely destroyed in a great storm in January 1917. After the death of her father, Ella and her sister Patience helped support her widowed mother and siblings by inshore fishing. On 8 September 1917, Ella was at sea fishing with her ten-year-old cousin, William Trout, when they heard a huge explosion. The Newcastle vessel SS *Newholme*, en route for the Tees with iron ore from Bilbao, had been torpedoed by the UC-31 about a mile off Start Point. As the wind and tide were against them, Ella and William lowered their sail and rowed as quickly as they could for the wreckage. Only a motor-powered boat reached the scene before them and, while William steadied the boat, Ella pulled a half-naked black sailor from the sea. He was in a bad way, but Ella wrapped him in a sail and resuscitated him. Eight other survivors were picked up by a motor boat, but Ella and William, after searching fruitlessly for further victims, reached a patrol vessel and handed over the man they had rescued. Afterwards, Ella was awarded the Order of the British Empire.[75]

Though the numbers of sinkings fell drastically in 1918, with only 71 first-class British fishing vessels working from ports in the British Isles succumbing to enemy attack, German U-boats managed to pull off a couple of further successful large-scale attacks, albeit away from the more heavily protected waters. On 30 May 1918, the *Sparkling Wave* was one of 11 fishing vessels from Northern Ireland which were collectively captured and sunk by the UB-64 while working around 26 miles west-north-west of the Calf of Man. The U-boat apparently surfaced in the middle of the fleet and ordered the crews to row for the shore in their punts. But one of the vessels, the *Never Can Tell*, apparently had no punt, and so the crew were ordered onto the deck of the submarine and plied with gin and cigarettes while they waited for the sinkings to be completed. Afterwards, the German commander put the fishermen onto a surviving boat and let them head for the shore.[76] The final mass sinking occurred in August 1918 in Canadian waters across the other side of the Atlantic, in the seas off Newfoundland and Nova Scotia. Thirteen vessels were lost. They proved to be the last such action during the war. Only two more fishing vessels were to be lost in the last couple of months before the Armistice and both were sunk by mines (Figure 10).[77]

---

75 British Parliamentary Papers, Fisheries England and Wales: Fisheries in the Great War: Being the Report on Sea Fisheries for the Years 1915, 1916, 1917 and 1918 of the Board of Agriculture and Fisheries, 50.

76 The Wreck Site, FV *Sparkling Wave*: https://www.wrecksite.eu/wreck.aspx?64812 (accessed 19 September 2018).

77 *British Vessels Lost at Sea, 1914–1918*, 124–125.

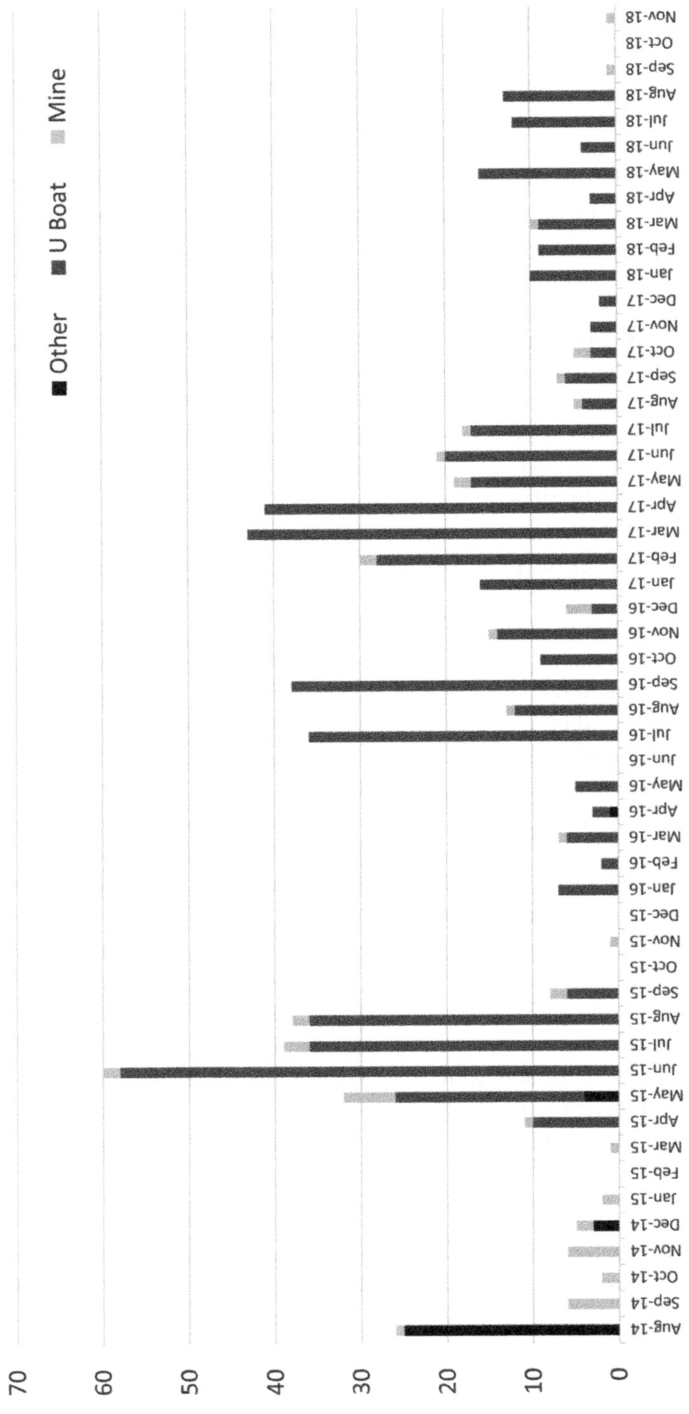

Figure 10: Fishing vessels lost to enemy action while fishing during the Great War, August 1914–November 1918
Source: HMSO, *British Vessels Lost at Sea, 1914–1918* (London, 1919).

Overall Impact

Not surprisingly, given the losses to enemy actions and the mobilisation for naval services of thousands of vessels and men, catches of fish landed in the United Kingdom declined dramatically during the Great War. Annual landings of catches in England and Wales, for example, for the period 1909–1913 inclusive, had averaged 14,451,200 cwt (735,000 tons), but for the years 1915–1918 the average landings fell to 4,690,000 cwt (240,000 tons). Given these figures, an impartial observer might have expected fish to be rationed, but this never happened and fish was not subject to price controls until March 1918. The excess of demand over the diminishing supply is reflected in the returns. Although only around a third as much fish was landed in England and Wales in 1918 as was landed in 1913 the value of the total catch almost doubled. Similar trends were evident in Scotland.[78] The average return per cwt (51 kg) was not only influenced by inflationary trends but also by shifts in the types of fish being caught. Fewer cheaper varieties such as herring were sought (much of which was in peacetime destined for the export market): a greater proportion was made up of more valuable demersal fish, preferred by domestic consumers.[79]

In its Report for 1920, the Fishery Board for England and Wales claimed that no group of workers had contributed more to the various dimensions of the war effort than fishermen, and certainly those still fishing seem to have performed well in relative terms. Although landings slumped in all belligerent countries, the British fishing industry seems to have managed to outperform its German counterpart: in 1917, for example, the British and German landings stood respectively at 31 per cent and 9 per cent of their 1913 figures. Though cut back to the bone, the British fishing industry and its workforce continued to make an invaluable contribution to the national food supply until the very end of the conflict.

---

[78] Robb Robinson, 'War, Public Intervention and Structural Changes in the British Fishing Industry, 1914–1930', in David J. Starkey and A. Tove Austbø (eds), *Maritime Industries and Public Intervention, The Fourth North Sea History Conference* (Stavanger, 2002), 126.

[79] Ibid., 126–127.

Chapter Seven

1918: Minesweeping and Anti-Submarine Operations during the Final Year

Minesweeping

By January 1918, the German minelaying assault on the waters around the British Isles and elsewhere had seemingly passed its peak. In part, this was probably due to the growing effectiveness of British countermeasures; and perhaps partly because it was increasingly difficult for the Germans to maintain such high levels of mine production, given the pressing demands of other military priorities. But while there were clear indications that fewer mines were being deployed than had been the case some six months earlier, an evident focus on more specific objectives can be discerned from a study of minelaying activities during the last year of the war.

Whilst the seaways in the Harwich area, as well as the waters adjacent to a few other British ports, continued to be strewn with mines deposited principally by Flanders Flotilla U-boats still operating from the Belgian ports of Zeebrugge and Ostend, the minelaying submarines of the High Seas Flotilla increasingly concentrated their efforts on three other specific dimensions of maritime activity considered to be of crucial strategic importance to Britain. First, the High Seas Flotilla pursued a programme of intensive minelaying off the coasts of the Netherlands, to disrupt and damage convoy traffic to and from the United Kingdom through these waters; these voyages were known to those involved as the 'beef trip'. The second area on which they invested a great deal of their effort encompassed the inner waters of the Firth of Forth and the tracks taken by the Scandinavian convoys. Finally, between the months of March and September 1918, the Germans expended considerable amounts of their precious time and scarce resources on a complex plan intended to finally ensnare the Grand Fleet. This involved the laying of a great barrier of mines out in the North Sea on a 45-mile radius from Bell Rock.[1]

Although there was a diminution in German minelaying off other coasts of the British Isles, such activities did not completely cease. On 21 February 1918,

[1] NA, ADM 186/604, 'History of British Minesweeping in the War', 70.

for example, a cluster of mines was deposited in the seas about one and a half to two and a half miles north-east of the River Tyne by the UC-49, presumably to try and sink a battleship returning northwards to its Scottish base after refit in north-east England. Although this field destroyed the fishing trawler *Reaper* and its crew,[2] eight mines were swept up and four others accounted for before further damage could be inflicted on shipping entering or leaving the Tyne.[3] After this, there was little further enemy minelaying activity in the waters off Durham, Northumberland, and the border regions until June 1918, when 21 more mines were reported to have been laid southwards of the war channel's junction with the Granton channels and, of these, 17 were soon accounted for by minesweeping forces.[4]

Whilst some areas, such as those swept by the Grimsby minesweepers, reported no new fields being laid during 1918, there was, as yet, no such respite in Harwich waters. Twenty enemy mines were destroyed there in January. A month later, 54 more mines were found near the Shipwash Lightvessel, while a couple of steamers were sunk by mines laid in the local stretch of the war channel when a decision was taken to allow ships to move through these waters to ease intense traffic pressures, even though minesweeping operations had been curtailed by a combination of poor visibility and bad weather. Similar pressures accounted for the sinking of three more steamers on 5 March, all lost within ten minutes and a half-mile radius of each other. A total of 36 mines were swept up during that month, many having been laid in the exit channel for the Harwich naval forces.[5] Attacks here increased again in April when 51 mines were accounted for, and two more minesweepers were lost, one of which was the requisitioned Belgian trawler *Numitor* and the other was the hired railway steamer *St Seiriol*,[6] which had done yeoman service and been associated with the destruction of more moored mines than any other British vessel.[7] Many of the experienced crew members lost with the ship were fishermen reservists.

On 10 May 1918, Vicco von der Luhe, commander of the UB-16, was found swimming near the Sunk Lightvessel. He turned out to be the only survivor of this veteran minelaying U-boat that had been a scourge of the East Anglian coasts since being commissioned back in May 1915, but had been eventually caught unawares on its old hunting grounds by a torpedo fired by the British submarine E-34.[8] Von der Luhe's war had ended but he never saw Germany again, succumbing to the influenza epidemic while still a prisoner in England the following March.

---

2 The Wreck Site, FV *Reaper*: https://www.wrecksite.eu/wreck.aspx?64064 (accessed 19 September 2018).

3 NA, ADM 186/604, 'History of British Minesweeping in the War', 74.

4 Ibid.

5 Ibid., 76.

6 Dorling, *Swept Channels*, 261–264.

7 NA, ADM 186/604, 'History of British Minesweeping in the War', 74.

8 Grant, *U-Boats Destroyed*, 122.

During May and June 1918, a further 29 mines were dealt with by Harwich minesweepers; and in August, another eight were found, presumed to have been laid with the intention of trying to catch a returning Dutch convoy. But this consignment seems to have been the final German throw of the dice here, and marked the end of their offensive in Harwich waters. Only seven old mines were located and destroyed during the last couple of months before the Armistice.[9] Yet the minesweeping forces from this coast were accorded little respite from their labours, being kept busy during the autumn by the pressing need to clear mines on the other side of the North Sea when the Belgian ports of Ostend and Zeebrugge fell back into allied hands between 17 and 19 October 1918. The minesweepers initially sent there tried to work under a hail of fire from the enemy's remaining coastal batteries, until these too were finally abandoned as allied land forces made further advances through Belgium.[10] Between 18 October and 6 November, 93 mines were located and destroyed off the Flanders coast, and 21 of these were found inside the Mole at Zeebrugge.[11]

The Great Barrage

The great barrage that the Germans tried to lay on one specific stretch of the Scottish coast during the spring and summer of 1918 was an undertaking on a considerable scale. The intention was to spread mines across these seas on a 45-mile radius from the Bell Rock, which lies 11 miles to the eastwards of the Firth of Tay. This German operation was a key part of an even wider strategy aimed at ensnaring the dreadnoughts of the Grand Fleet amongst these mines, should they sail southwards from their Scapa base in order to counter perhaps a final foray by German High Seas Fleet battleships heading across the North Sea and towards Britain.[12] To create this elaborate trap, German minelaying submarines toiled away from April to August, criss-crossing the North Sea to deposit cargo after cargo of deadly devices.[13]

By any naval measure this was certainly a vast scheme. But, not long after it commenced, British sweepers began locating groups of these mines and the overall pattern behind the German's intention was soon discerned and these findings kept quiet. There was no publicity, the Germans were given no intimation that their plans had been compromised, and, as each group of mines was laid, they were soon sought out and swept up by British minesweepers. The Germans laboured here in vain. Allowing them to continue with their efforts to

9 NA, ADM 186/604, 'History of British Minesweeping in the War', 76–77.

10 Ibid.

11 Ibid.

12 Friedman, *Fighting the Great War at Sea*, 340.

13 Hurd, *History of the Great War*, 3:261.

lay great numbers of mines in this region meant that other sea areas and shipping were spared. But thwarting a plan of this scale required considerable resources and a substantial number of minesweepers and crews were transferred to the Firth of Forth area from other regions. By the end of October more than 60 vessels were engaged in removing the mines.[14]

The British secret was well kept; and the German naval authorities thought their barrier was intact until after the Armistice. Yet, even though they never realised their minelaying activities had come to nought there was a further reason why the Germans never got the opportunity to trigger their grand design. Their plans for the High Seas Fleet to make a final foray – and thus lure out the Grand Fleet – were scuppered when German naval crews mutinied on the night of 29 October as their battleships readied for sea.[15]

Beyond the British Isles

The Auxiliary Patrol minesweeping vessels and their predominantly fishermen crews stationed in the Mediterranean continued to have their work cut out almost to the last days of the conflict. When the British army advanced into Palestine and a base was established at Haifa at the end of September 1918, five trawlers and three motor launches were sent to the port and disposed of 57 mines while clearing the channel and anchorage.[16] The movement of the army under General Allenby northwards to take Damascus at the beginning of October created yet more work for the minesweepers, and after Haifa was cleared the Auxiliary Patrol vessels moved up the coast to sweep the approaches to Beirut. The Turkish situation in south-eastern Europe deteriorated rapidly and when Ottoman forces fell back as the British advanced in Syria their government sought an end to hostilities. After a period of negotiations, the Armistice of Mudros was finally concluded on 30 October, but, anticipating the Turkish capitulation, the British had already begun assembling a minesweeping force at the port of that name on the island of Lemnos in the northern part of the Aegean Sea. On 1 November, just two days after the Turkish Armistice was signed, this force, consisting of a kite balloon ship, five twin-screw steamers, 14 trawlers, eight drifters, and three motor launches and boasting a large contingent of fishermen reservists amongst its crews, began to sweep a way through this substantial barrier of mines, which had proved such a formidable obstacle during the fateful Gallipoli campaign back in 1915. Though sweeping operations were rendered particularly difficult thanks to the shallow setting of some of the minefields, a swept channel was created and buoyed in just nine days. A total of 184 allied and 424 enemy mines were destroyed in the

14 Chatterton, *Auxiliary Patrol*, 253.

15 Ibid., 254.

16 Ibid., 302.

process, and afterwards a line of allied warships, led by HMS *Superb*, steamed up the channel to anchor off Constantinople, then Turkey's capital city.[17]

Far to the north, the forces left in the White Sea region were in an increasingly isolated and uncertain position. By early 1918, the allies were more concerned about what would happen to the stocks of war materials and other supplies stockpiled in north Russia, than with providing new consignments, and as the Bolsheviks moved towards an Armistice with Germany the situation ashore for the remaining British naval forces was ever more ambiguous and unfriendly. Eight minesweeping trawlers under the command of Lieutenant Ryan left the region bound for Lerwick in early February, carrying a range of British and Romanian service people, but other vessels apparently remained for some time after this.[18] Most of the serviceable trawlers and drifters left in the region seem to have departed by the beginning of spring. Allied forces were to return to Murmansk later in the summer in support of anti-Bolshevik forces, but this was to involve them in hostilities which were to outlast the Great War.[19]

Extensions to the War Channels

Meanwhile, back in home waters, during the course of 1918, the east coast war channel had been extended and buoyed as far north as May Island in the Firth of Forth, and the new sections were kept clear of obstructions by minesweepers. This meant that Norwegian-bound convoys, which by then utilised the Fife port of Methil as a staging post, could now follow swept channels for their entire coastal passage. A similar buoyed war channel was also laid along the south coast of England, from Dover as far westwards as Anvil Point in Dorset where the passage then connected with the Portland Eastern and Southern Swept Channels. The hazardous routeway by St Catherine's Point on the south coast of the Isle of Wight was avoided by routing the channel through the Solent. Protected anchorages for merchant vessels were also created near to Selsey Bill in West Sussex and in the waters off Torquay. The route was not buoyed to the west of Portland but hugged the shore. The use of such shallow waters helped to deter U-boats from making torpedo attacks, but still required the extensive use of minesweepers to counter any threat of renewed German mining operations.[20]

Although the number of vessels lost to mines fell quite dramatically during 1918, the situation at sea remained extremely serious for many months to come.

---

17 NA, ADM 186/604, 'History of British Minesweeping in the War', 82.

18 NA, ADM 137/883, White Sea Telegrams, 2 February and 27 February 1918.

19 See, for example, Erik Goldstein, *The First World War Peace Settlements* (London, 2002); Clifford Kinvig, *Churchill's Crusade: The British Invasion of Russia, 1918–1919* (London, 2006); Ian C. D. Moffat, *The Allied Intervention in Russia, 1918–1920: The Diplomacy of Chaos* (Basingstoke, 2015).

20 NA, ADM 186/604, 'History of British Minesweeping in the War', 83–84.

According to Hurd's figures, the average monthly losses of merchant ships during the period February to December 1917 inclusive was 104 vessels, while the corresponding average monthly losses from January to October 1918 inclusive was 54 vessels.[21] Thus, although the monthly toll of shipping losses never again approached the peaks experienced in the first half of 1917, it remained uncomfortably high. Attacks by U-boats, using guns or torpedoes rather than by laying mines, continued to be the principal cause of mercantile shipping losses, and during this last year of the war the fishermen and their armed trawlers and drifters serving with the Auxiliary Patrol on anti-submarine duties remained on the front line of the maritime fight against Germany.

Turning the Tide

By early January 1918, however, the Auxiliary Patrol vessels and crews engaged in this war were much better equipped for this fight than they had been back in 1914. Not only were their fishermen crews now better trained, and vastly more experienced in maritime war than they had been in the early months of hostilities, they were also much better supported. The growing expertise of fishermen in terms of the war effort had been marked by alterations in the RNR ranks. Initially, the rank of skipper had been created back in 1911 for the RNR(T), and many masters of fishing vessels served as either skippers or temporary skippers for the duration of the war. But in 1917 a new rank, that of Chief Skipper, was created, and one of the earliest fishermen to be promoted to such a position was Peter Yorston, the Aberdeen skipper who had also been the first skipper appointed to the RNR(T) when it was created in 1911.[22] Only a relatively small number were raised to this rank, with just 60 chief skippers and 120 temporary chief skippers being appointed during the conflict, but a number of these men were put in charge of units or groups of vessels.[23] Such roles had in the early stages of the conflict often been the preserve of retired naval officers or senior grades from the mainstream RNR or RNVR. Typical of those who stepped up to the new rank was the Hull skipper Charles Allum. He was one of those trawlermen taken into the RNR(T) in the early weeks of the war and initially appointed temporary skipper, later skipper, of the armed trawler *Viola* and had served as part of an anti-submarine unit patrolling from Shetland until October 1916, when his ship and unit were transferred southwards to the Tyne. He was later mentioned in despatches for two incidents in 1917.[24] Allum was amongst a number of skippers who were promoted to the new rank in January 1918, and

---

21 Hurd, *History of the Great War*, 3:379.

22 Eyre, *Skippers of the Royal Naval Reserve in the First World War*, 642.

23 Ibid., 23.

24 Robinson and Hart, *Viola*, 108–109.

took charge of his section of trawlers, his service record later stating that he had carried out this role with 'zeal and ability'.[25]

For the skippers and crews of Auxiliary Patrol vessels, much of the conflict, in the North Sea at least, was now concentrated in the east coast war channels, and naval vessels there were by now working closely with aircraft and airships in carrying the war back to the U-boats. A not untypical example of such co-operation took place on the east coast on 25 April when a Sopwith Baby biplane first spotted the wake of a submarine in the vicinity of the war channel off Filey Bay. Once located from the air, the German vessel was then attacked on the surface by the armed trawler *Sparrow*, which opened fire with its howitzer.[26] The submarine was not sunk, but probably disrupted in its intent. Airborne patrols were by now often quite thorough, sometimes of considerable duration by contemporary standards. One aircraft, made by Short, when operating from Killingholme on the Humber, was reported to have made a patrol flight lasting six hours and ten minutes, afterwards described as a 'very creditable performance'.[27] The continued use of the convoy system throughout 1918 also, of course, meant that German submarines now often had little option but to attack merchant vessels provided with escorts, rather than preying on individual ships devoid of much protection, and many merchant ships were also now fitted with guns. Additionally, the use of mined nets and other equipment used in this coastal maritime war continued to be refined and made more effective during that final year of hostilities.

Additionally, the Auxiliary Patrol vessels were themselves much better armed, and deployed more effective weaponry than had been available when many of the fishermen had initially signed up. By early 1918, most of the armed trawlers in the Auxiliary Patrol had been fitted with 12-pounder guns, or even, like the armed trawler *Sparrow*, with a howitzer. Many of the drifters now also carried 6-pounder guns. In addition, fishing vessels, now grouped into the fishery reserve, were also provided with supplementary armed cover. The several types of hydrophone that had been developed and manufactured were by now distributed widely amongst the anti-submarine vessels. Auxiliary Patrol crews had regular opportunities to deploy these listening devices on active service and the lessons learned as operational experiences accumulated increased their effectiveness.

Another particularly crucial piece of kit was the depth charge, which prior to 1918 had only been available in very limited quantities. When first developed, as we have noted, only two depth charges had been available per ship, but by early 1918, with large-scale production now under way, between 30 and 50 such weapons were allocated to bigger anti-submarine vessels, most notably destroyers

25 NA, BT/377/7, WSA 260, Charles Allum RNR Service Record.

26 NA, ADM 137/721, Auxiliary Patrol Humber, 25 April 1918.

27 Ibid., 2 March 1918.

and the larger trawlers. From at least the middle of February, the Admiralty was advocating the liberal use of depth charges by Auxiliary Patrol vessels.[28] New methods of propelling them even further from a vessel's side were also introduced.[29] By the end of the war, at least 285 armed trawlers had been issued with depth charges.[30]

The outcome of these developments meant that the U-boat tactic of submerging, of slipping beneath the waves when pursued, no longer always provided such a safe haven, for not only could a submarine now often be tracked with hydrophones but the seabed on which it sought refuge could be subjected to a widespread and sustained depth charge assault. A survey of the 1918 Admiralty reports from North Sea coasts of Britain suggests that the Auxiliary Patrol armed fishing vessels and other naval units were picking up and thwarting far more U-boat incursions than had been the case in the first years of the conflict. The drifters of the Dover Patrol, for example, were certainly, as Chatterton puts it, 'more dangerously active' than ever. This was through working closely with the new Dover Deep Mine Barrage that had been first proposed by Admiral Bacon when Commander of the Dover Patrol in early 1917. The barrage proved a far more elaborate, ambitious, and effective attempt to block the English Channel to U-boats than its predecessors, all of which were intended to force U-boats to take the longer voyage around the North of Scotland. Much of this new barrier was laid during November and December 1917, and by early 1918 it stretched from Folkestone in England to Cape Gris Nez in France: something like six or seven miles in width. The barrage consisted of staggered rows of the newer and more effective British contact mines, largely copies of captured German devices, that were laid at varying depths thus making a passage below the surface an extremely hazardous affair. Any U-boat seeking to make a crossing therefore needed to try and slip through on the surface at night, but to do this they had to run the gauntlet of a panoply of British forces. To deter surface crossings of the barrage, the British deployed anti-submarine nets and set up a system of constant patrols by destroyers, motor launches, armed trawlers, and drifters, while airships and aeroplanes overflew the waters. After dusk fell, the waters were illuminated by vessels carrying searchlights or armed trawlers burning 'Dover Flares' – powerful lights which lit up great expanses of the sea. Any U-boats spotted on the surface were faced with the option of engaging with these armed ships or else diving and chancing their luck amongst the thick forest of mines that were laid below. Best figures available suggest that at least ten U-boats were lost while trying to cross this barrage, and Auxiliary Patrol vessels were involved in a number of these sinkings, the first of which was UB-56, which was

---

28 ADM 137/804 16 February 1918.

29 Fraser M. McKee, 'An Explosive Charge', *The Northern Mariner/Le Marin du nord* 3.1 (1993), 52–53.

30 Ibid.

destroyed even before the barrage was fully completed, disappearing in a terrible explosion on the evening of 19–20 December 1917, around 11 miles south-east of Folkestone.[31]

The way Auxiliary Patrol vessels worked in combination with the other elements of the barrage is exemplified by the destruction of the UB-38 during the night of 8 February 1918. The U-boat had been attempting to pass over the barrage and slip by the British patrols when it was spotted by the drifter *Gowan* II after being illuminated in the flare of one of the armed trawlers. The *Gowan* II chased at full speed, and, after a pursuit of 20 minutes, the submarine was left with little option but to dive and then be blown up amongst the mines. Just over a month later, on the night of 10 March, the UB-58 was similarly spotted by a division of five drifters when illuminated in the flare of the armed trawler *Jaboo* II, while also attempting a transit of the barrier just over six miles south-south-west of Folkestone, and forced to dive to its doom as they closed in. On 11 March, another six drifters, *Ocean Roamer*, *Kessingland*, *Loyal Friend*, *Herring Seeker*, *Harvest Hope*, and *Lord Leitrim*, along with the armed trawler *Fuji*, forced the UB-33 to dive to its destruction in much the same area, leaving behind only oil, assorted wreckage, and a dense cloud of smoke.[32] The barrage and patrol vessels had a further success on 22 April when the drifter *Shipmates*, patrolling in the vicinity of the Varne Buoy, was shaken by a heavy explosion while searching the vicinity with the drifter *Seaflower*. She rescued six German submariners from the sea, who told of their vessel's fate. Their ship, the UB-55, had left Zeebrugge the previous evening then voyaged in darkness on the surface until having to dive after catching sight of a patrolling drifter, where she was destroyed after fouling a mine. The need to dive amongst the mines to avoid the drifter patrols was possibly also the cause of the losses of the UB-31, UB-78, and the UC-79 amongst the barrage during April and May 1918. In June, the UC-64 also succumbed amongst the mine barrier, at a place quite close to the Varne, after being forced to dive when encountered by yet another group of seven drifters patrolling together with one trawler.[33]

Enemy Attacks on Dover Patrol Operations during 1918

Well before these later U-boat losses, the Germans had sought to destroy barrage operations by making yet another surface raid on Dover Patrol vessels from their Belgian coast bases, which took place on the night of 14–15 February 1918. That evening, 58 Dover drifters, split into nine divisions, were out patrolling along the Folkestone to Cape Gris–Nez barrage, in company with a couple of motor

31 Gibson and Prendergast, *German Submarine War*, 231.

32 Chatterton, *Auxiliary Patrol*, 261.

33 Gibson and Prendergast, *German Submarine War*, 311.

launches, while several other armed trawlers and paddle minesweepers and a couple of destroyers were also sailing in the vicinity, many providing illumination to both the north-west and south-east of the Varne Shoal.

A group of four German destroyers made the initial attack, first circling and shelling the paddler *Newbury* from both sides, before steaming slowly down a line of drifters, using their searchlights to illuminate and bombard each vessel in turn. Although the drifters *Veracity* and *Tessie* were sunk during the onslaught, many others escaped under a heavy hail of fire. The paddler *Lingfield* and a motor launch which came up to investigate the firing were less fortunate. As this first group of German warships made off, another three destroyers joined in the fray and attacked another group of patrol vessels, starting with the trawler *James Pond*, one of the new Admiralty class armed trawlers which was acting as an illuminating vessel. The *James Pond* was commanded by Skipper Alfred Edward Berry, who had already won a DSC while minesweeping in the Dardanelles back in 1915. Berry, who was one of those fishermen who had now been promoted to the rank of Chief Skipper, was badly wounded when a shell entered his cabin during the attack, but he managed to maintain control and tried to put out the ship's fires before eventually giving the order to abandon ship. A few minutes afterwards the *James Pond* blew up. The survivors eventually reached the French coast and safety at Wissant.[34]

Whilst the *James Pond* burned, the German destroyers continued their attack, moving north-west along the line of vessels on the barrage, sinking the drifters *Clover Bank*, *Comet*, and *Jeannie Murray* while causing serious damage to five more vessels. The *Violet May* was one of the drifters attacked, and while the enginemen were giving the vessel all possible steam the first two German salvos destroyed the wheelhouse, killing all on board other than the two engineers, James Ewing and Alexander Noble, and the mate and a deckhand, both of whom were badly injured. Another shell struck the boiler, which burst, while the two enginemen managed to launch the open boat after clearing it away from the boom and other debris. They hauled the two injured crewmen on board and rowed away but, seeing the drifter did not sink, they returned on board, struggling amongst the dead bodies of their shipmates to douse the flames with buckets of seawater while ammunition from the small arms kept in the wrecked and still burning wheelhouse went off around them. They managed to bring the fires under control then lifted their injured colleagues back on board and made makeshift flares out of old clothing dipped in paraffin.[35] After a few anxious and uncertain hours they were located by the drifter *Courage* and towed into Dover. The mate later died from his wounds, but Ewing and Noble were both awarded the Conspicuous Gallantry Medal for their actions that night.[36]

---

34 Chatterton, *Auxiliary Patrol*, 199.

35 *Daily Record*, 18 March 1918.

36 *London Gazette*, 15 March 1918.

A roll call of those lost in the attack on the *Violet May* provides some indication of the way fishermen from so many different coastal communities were drawn together to fight in this war: Skipper Alexander Buchan and trimmer Paul H. Gerrie hailed from Fraserburgh. Fellow trimmer John Henry Erskine came from Blyth, deckhand Donald Gillies's parents lived in Strond on the Isle of Harris, while the parents of deckhand W. J. Harris lived on the south coast in Folkestone. The youngest member of the crew, 17-year-old cook James Dyker, was an Aberdonian, while the mate, Charles Barber, and deckhand, Albert Allen, came from Lowestoft.[37] Like many of other fishermen from the Auxiliary Patrol whose bodies were recovered after such actions, the remains of most of the *Violet May*'s crew lie individually in many different cemeteries, often close to their old homes, in graves up and down the length of the coasts of the British Isles.

Further vessels were attacked before the enemy destroyers made off into the darkness. Later in the morning the surviving drifters came back into Dover and landed their dead and injured crewmen, Crowds gathered as the injured were transported to the hospital. The dead, wrapped in blankets, were placed on stretchers and carried into the Market Hall. One trawler and seven drifters had been sunk during the raid, while a paddle steamer and six drifters were damaged. Altogether some 89 officers and men, many of them RNR fishermen reservists, lost their lives.[38]

The following night a German submarine bombarded Dover, killing a child and injuring several townsfolk, but their efforts to destroy the barrage had failed.[39] The armed drifters and trawlers soon returned to their duties on the barrage and the U-boat losses continued.

Western Waters

In April 1918, as more effective mines were becoming available in substantial numbers, the British and the Americans began laying a deep minefield barrage to replace the existing line of drifters and nets guarding the North Channel between Scotland and the north of Ireland. The strategic importance of these waters had increased because these seaways were by now being used to bring large numbers of US troops for disembarkation in either Liverpool or Glasgow. Sometimes several U-boats were reported to be lying in wait for the large troopships to arrive and U-boats sightings and attacks had been reported as such traffic built up.

In late January 1918, the RMS *Andania*, a veteran of the Gallipoli troop landings, was torpedoed off Altacarry Head in Northern Ireland by the U-46, when outward bound for New York from Liverpool. Several Auxiliary Patrol

37 Commonwealth War Graves Commission Records.

38 Ibid.

39 Chatterton, *Auxiliary Patrol*, 203.

vessels went to its assistance, passengers transferred, and the ship was taken in tow. During the attempted salvage the periscope of a U-boat was spotted, presumably trying to manoeuvre to deal a final blow, but the *Davara* fired two shots at almost point-blank range, the shots falling just too short and slightly off target. The *Cormorant* II, also near, put its helm over and fired one shot as it began a ramming run. The submarine disappeared under the bow and while the engineman rushed on deck to report the hull grating over something solid the trawlers fired a depth charge, as did the incoming drifter *Young Fred.* The U-boat was defeated in its plan, but the attempts to save the *Andania* were not successful and the vessel sank a few hours later off Rathlin Island. All passengers appear to have been saved but seven crew were lost with the ship.[40]

The attacks and sinkings continued. At 3.30 p.m. on 5 February 1918, the SS *Almanace* was torpedoed off the Maidens while en route from Baltimore to Liverpool with general cargo and every available trawler, drifter, and motor launch in Larne Harbour went to its assistance. The vessel sank at 4.45.[41] At 6.26 on the same day, the UB-77 succeeded in torpedoing the *Tuscania*, an Anchor Line steamer carrying 2,235 troops. The attack took place on a bitterly cold night, some seven miles west of Islay, and, as the stricken ship listed to starboard, it proved almost impossible to lower the port side lifeboats. Many of those who got into other lifeboats came ashore under the 660 foot high rocky cliffs of the Mull of Oa, Islay's most southerly point. Locals helped rescue survivors dashed onto the shore, and the island community provided warmth and sustenance in their houses and hotels. A total of 166 soldiers as well as several British sailors lost their lives in the sinking. Many of the American troops hailed from states such as Idaho, long distances from the sea. Royal Navy vessels, including several destroyers and the largely fishermen crews of a number of the Auxiliary Patrol armed vessels, including the trawlers, *Elf King*, *Cardiff Castle*, *Corry Roy*, *Walpole*, *Belona*, *Gloria*, and *Samuel Barkas*, also played a major part in rescuing those who took to the waves.[42] In October 1918, just a few weeks before the Armistice, another ship, the *Otranto*, full of American troops, sank off Islay, this time as the result of a collision, and hundreds more were drowned. Today, both tragedies are commemorated by the American Monument, built on the high cliffs of the island's Oa Peninsula.[43]

Although the loss of the *Tuscania* spurred forward these plans to improve the sea defences between Ireland and Scotland, the intended minefield of the North Channel Barrage was never fully completed, even though two lines of hydrophone trawlers were already in place by the opening of 1918. One of these

---

40 NA, ADM 137/808 Larne, w/e 27 January 1918.

41 Ibid., w/e 9 February 1918.

42 Ibid.

43 'Welcome to Islay Info': https://www.islayinfo.com/lord_robertson_islay_troopships.html (accessed 19 September 2018).

lines stretched across the mouth of the Clyde, while the other reached across from the Mull of Galloway on the Scottish side to Skulmart, to the south of Belfast.[44]

During the crucial spring and early summer of 1918, as the transatlantic troopship traffic intensified, the drifters patrolling near the North Channel had some success in harassing and hunting the German interlopers. On the evening of 17 April, the drifter *Pilot Me*, while on duty in these waters, spotted a periscope close on the starboard quarter, turned around and went for it, dropping a string of four depth charges in succession while zigzagging over the spot before lowering hydrophones to listen for movement. No obvious sounds were picked up, but the damage had been done. About 20 minutes after the first depth charge had detonated, the UB-82 broke surface between the *Pilot Me* and another drifter, the *Young Fred*, bow first and at a very steep angle. The various drifters which had by now gathered opened fire at very short range. Meanwhile, the *Young Fred*, which was no more than 100 yards away from the submarine when it broke surface, turned around and steamed straight towards the German craft. The submarine disappeared stern first under the *Young Fred*'s bows and the drifter crew dropped two D-type depth charges right on top. The depth charges blew up below and were followed by another and reportedly distinct explosion of much greater proportion, and a huge column of water erupted from the sea. When everything subsided, the sea was covered with oil and pieces of wreckage, and various articles were picked up including two German seaman caps, internal woodwork fittings, bedding, and two postcards. There were no survivors.[45]

A fortnight later, on 30 April, the Larne drifter *Coreopsis* sighted the conning tower of the UB-85 in the North Channel just before 3 a.m. The commander, Lieutenant Percy Sutcliffe Peat, summoned his gunner called MacDonald, who, in his haste, tumbled onto the deck clad only in his shirt. The *Coreopsis* attacked at full speed and MacDonald fired its 6-pounder gun at close range as the U-boat ran across the drifter's bows, the initial shot reportedly hitting the conning tower. The submarine then made off, and the much slower drifter followed in what seemed like a vain pursuit. Then, suddenly, the German vessel altered course, hove to, and fired a Very light, while its crew shouted that they were surrendering. Peat was suspicious: the U-boat was more heavily armed than his little drifter, so he manoeuvred until he had placed the submarine in the moon's rays and ordered several rounds to be fired at its gun.[46] There was no response, so the *Coreopsis* ceased fire but kept a close watch on the U-boat until other patrol vessels approached, after which Lieutenant Peat and his skipper George Ernest Stubbs sent their open boat towards the Germans and told them to jump into the sea where they would be picked up.[47]

---

44 Hurd, *History of the Great War*, 3:66.

45 NA, ADM 137/808 Auxiliary Patrol Larne, 20 March 1918.

46 Ibid.

47 Gibson and Prendergast, *German Submarine War*, 302.

Four boatloads of prisoners were eventually brought on board *Coreopsis* and the HM Drifter *Valorous* sent its boat across a couple of times as well. All prisoners were battened down in the hold and the *Coreopsis* was escorted into Larne by the armed trawlers *Lewis Reeves* and *William Symons*.[48] The whole incident lasted little more than 90 minutes.

Later, the reasons for what seemed at first sight a remarkably easy surrender became more evident. Patrol craft had forced the U-boat to remain continually submerged for long periods as it approached this sea area, and the on-board batteries had given trouble, the chlorine gas making many of the crew quite ill. The final straw came when one of *Coreopsis*'s shells damaged a hatch. Fifteen tons of water then poured into the conning tower, cut off the passage to the magazine and got amongst the batteries, almost asphyxiating the crew. Later, Lieutenant Peat was awarded the DSO, while his second in command, Skipper George Stubbs RNR, a Gorleston fisherman in peacetime, received the DSC.[49] The Admiralty also awarded £1,000 which was divided amongst the crew.[50]

The Otranto Barrage in 1918

In early 1918, Auxiliary Patrol trawlers, drifters, and motor launches, by now backed by light destroyers and aircraft used for observation, were still very much on the front line in a range of anti-submarine activities, which included hydrophone patrols and mobile and fixed net barrages. There was considerable investment in hydrophone equipment and training in this last year of the war for the forces involved in the Otranto Barrage.[51] In April 1918, a final fixed and extensive new deep barrage of mined nets began to be laid by allied forces in the Straits of Otranto (Photographs 9 and 10). This was an ambitious affair and intended to extend from the eastern end of a deep minefield laid by allied forces a few miles from Otranto across the Straits towards the islet of Fano, part of the Ionian archipelago. Although not completed until the end of September, only six weeks or so before the Armistice, its construction was evidently perceived by the German and Austrian submarines to be causing problems as early as June.[52] On 10 June, Austria sent two battleships down the Adriatic to disrupt barrage operations and attack the patrol vessels, but they were sighted by two Italian motor launches which, despite the marked disparities in size and armament, attacked with torpedoes and actually sunk the Austrian dreadnought

48 NA, ADM 137/ 808, Auxiliary Patrol Larne, 20 May 1918. See also *Edinburgh Gazette*, 20 November 1918.

49 Fifth Supplement to the *London Gazette*, 4 June 1918.

50 Hurd, *History of the Great War*, 3:257.

51 NA, ADM 137/1579, Otranto Barrage, 26 June 1918.

52 Chatterton, *Auxiliary Patrol*, 210–211.

Photograph 9: Armed trawler in Italy, 1918
Source: Joy Simmonds

Photograph 10: Armed trawler crew in Italy, 1918
Source: Joy Simmonds

Szent Istvan, thus thwarting what could have been a serious attack on the Auxiliary Patrol forces working on the new barrage.[53]

Although the new Adriatic barrage and associated Auxiliary Patrol operations were not as effective in terms of curtailing U-boat movements as their counterparts in the Dover Straits during that final year of the war, they did bring about the destruction of the UB-53 in early August 1918.[54] This U-boat was heading for the Mediterranean from its base at Pola, but after diving deep under the Fano side of the barrage became entangled with the final section of the fixed mine net line. Two violent explosions badly damaged the U-boat which lost its trim and struggled to the surface. Unable to dive again, the submarine commander sighted a couple of armed trawlers and a destroyer closing from the eastwards. Escape seemed impossible, so he ordered the crew into the water and scuttled the vessel. Several hours later, 27 submariners were picked up by the British destroyer HMS *Martin* and the trawler *Whitby Abbey*. On board the trawler, the C-tube hydrophone operator began picking up fearful banging and crashing sounds on his equipment, which seemed to grow louder and louder. Then, suddenly the face of a man appeared over the gunwale in front of him. It was one of the German submariners who had clambered aboard by way of the C-tube.[55]

The Auxiliary Patrol, Convoys, and Combined Operations in 1918

By 1918, Auxiliary Patrol vessels working from other bases around the British Isles were also demonstrating the benefits of better and more armaments and the lessons they had learned from more than three and a half years of war. Hydrophone hunting flotillas were now a regular feature of North Sea patrol work and the greater availability of depth charges was evident from the willingness of patrol vessels to use more and more of them when attacking U-boats as the year wore on.

Substantial numbers of Auxiliary Patrol armed trawlers and drifters based between the Humber and the Firth of Forth were deployed in the escort of convoys moving up and down Britain's east coast. Many of the actions that took place off these coasts during 1918 involved Auxiliary Patrol vessels and Royal Navy destroyers taking on U-boats which had launched attacks on the convoys. Many such assaults were driven off or deterred, although not always without loss. On 25 March, for example, the Armed Drifter, *Border Lads*, which had fished out of Eyemouth in peacetime, was sunk just two miles off the mouth of the Tyne by

53 Gibson and Prendergast, *German Submarine War*, 271–272.

54 Halpern, *Otranto*, 144.

55 Gibson and Prendergast, *German Submarine War*, 272–273.

the UB-78 while supporting a convoy.[56] A few days earlier this U-boat had sunk the steamer *Polleon* in the same vicinity, and a couple of days after sinking the drifter it made a further attack near the Tyne, this time on the tanker *British Star*, which, although torpedoed, made it back into port. A few weeks later, as noted above, the UB-78 itself fell victim to the Dover Barrage.

The stretches of seaway where convoys arrived or departed from ports were, unsurprisingly, often a focal point for action with the enemy. On the night of 17 March, for example, the 21-year-old Armed Trawler *Swallow*, which worked out of Hull in peacetime, sighted and engaged a submarine when about four miles eastwards of the Tyne while awaiting the arrival of a convoy. One round was fired as the submarine made off into the darkness, but the pursuit of the vessel was continued by motor launches and other hydrophone listening vessels. Though it seemingly escaped, the U-boat was evidently thwarted in its immediate plan to attack this convoy. The following day, another Tyne vessel, the Auxiliary Patrol Armed Yacht *Miranda* II, while accompanying a convoy towards Scarborough, spotted yet another submarine as it voyaged along the North Yorkshire coast and ran down its track, first dropping a depth charge and then remaining in the vicinity on guard while the convoy moved safely on its way. A hydrophone, hunting flotilla, other armed trawlers, and the destroyer HMS *Talisman* then joined in the search which continued for 18 hours, and there were further similar extended hunts during the same week.[57] Although no U-boats were sunk in these encounters, the level of harrying they encountered made it more difficult to carry out effective attacks on convoys.

But these were bruising and often fatal encounters for both sides. The Armed Trawler *Swallow* finally met its end a week or so later when it was run down by the SS *Audax* while hunting another U-boat off Skinningrove. The trawler sank in seven minutes, but all the crew were rescued.[58] The SS *Audax* did not survive the war, being torpedoed along the same stretch of coast by the UB-80 on 6 September.[59]

Many more pitched battles were fought with U-boats along the east coast of Britain during 1918. One typical encounter took place off the north-east coast of Scotland on 27 June, soon after the fishing drifter *Kingfisher* reported the sighting of a submarine to HM Trawler *Candidate*, which in peacetime fished out of Leith. As the armed trawler ran towards the sighting it spotted a periscope at 7.10 a.m., some 200 yards off its port bow, and a thin trail of paraffin along the surface of the sea. Two minutes later, *Candidate* was on top of the trail, dropping two depth charges. After the second detonation, a large quantity of oil

---

56 NA, ADM 137/804, Auxiliary Patrol Tyne, w/e 30 March 1918.

57 Ibid., w/e 23 March 1918.

58 Ibid., Auxiliary Patrol Tyne, w/e 30 March 1918.

59 *British Vessels Lost at Sea, 1914–1918*, 96 and The Wreck Site, *SS* Audax, https://www.wrecksite.eu/wreck.aspx?10929 (accessed 19 September 2018).

and bubbles broke the surface of the sea. A hydrophone was put over the side, but only the noise of hissing steam could be heard. At 7.25, the armed trawler *Transvaal*, which worked out of Fleetwood in peacetime, arrived and buoyed the position. During the afternoon, the HM Yacht *Nairn* and the Grimsby-owned Armed Trawler *Ida Adams* joined in the hydrophone watch. Nothing further was heard, and two further depth charges were dropped at 8.45 p.m., after which yet more oil came to the surface. A couple more depth charges were dropped the following day but without much result.[60] The U-boat had no doubt got away but had probably been subject to a terrifying ordeal.

Aircraft and airships were by now involved in a number of these actions and North Sea anti-submarine hunts near the east coast war channels became increasingly sophisticated as the year wore on. One such pursuit took place on 18 June when an aeroplane from Redcar sighted the periscope of a submarine off the Yorkshire coast at 5.10 in the afternoon. The destroyer HMS *Star* and HM Whaler *Arctic Whaler*, escorting a convoy, ran down on the spot and dropped depth charges, and then another aircraft guided the Armed Yacht *Miranda* II to the place where it had sighted oil on the surface. The yacht dropped a depth charge then listened on hydrophones and picking up the noise of a submarine to the eastwards of the spot, buoyed the position, and also observed oil. Sometime later several armed trawlers and drifters, including the *Bombardier*, *Viola*, and *Faithful Friend* arrived at the location and began listening with their hydrophones. Further depth charges were pitched into the sea, the destroyer HMS *Mohawk* joined the hunt, and a large upheaval of water was observed. The pursuit was maintained: an aeroplane from Seaton Carew came in and dropped a bomb on the thick oil and bubbles and the trawlers continued to listen on their hydrophones, finally picking up faint submarine noises around midnight. Although these later completely faded away, the hydrophone watches and aircraft patrols continued throughout the next day. At 5.30 p.m. on the 19th, the Armed Trawler *Viola* spotted a red signal light dropped by an aircraft and on reaching the place glimpsed a periscope, dropped three depth charges on the position, and then listened on hydrophones. Nothing further was picked up and although no sinking was confirmed[61] it was clear that the erstwhile U-boat predators were now themselves being subjected to relentless and quite complex hunting tactics.

Nonetheless, in the confusion of such actions there were sometimes misperceptions over submarine kills. The UB-107, for example, was long believed to have been sunk in a depth charge attack by the Armed Trawler *Calvia* and Yacht *Vanessa* II on 27 June, but whatever U-boat was attacked in that incident it is now thought not to have been this vessel.[62] The UB-107 is now believed to have torpedoed two steamers, *Chloris* and *John Rettig*, part of convoy HZ 45

---

60 NA, ADM 137/806 Auxiliary Patrol Peterhead, w/e 29 June 1918.

61 NA, ADM 137/805 Auxiliary Patrol Tyne, 18 June 1918.

62 Chatterton, *Auxiliary Patrol*, 298–299.

from Norway, a couple of hours after the time of this attack at a location much further to the south along the Yorkshire coast. UB-107 certainly disappeared: its own wreckage was discovered decades later, in 1985, off Flamborough Head, many miles to the south of its original supposed sinking place, lying beneath the remains of the steamer *Malvina*, which was reportedly sunk by the UB-104 on 3 August.[63] The causes of its loss remain unclear, but it possibly struck a British mine.

The Final Months

Such combat along the east coast continued throughout the remainder of the summer. For example, Auxiliary Patrol forces working out of the Tyne were responsible for the destruction of the UB-30, which had left its home base on 6 August intent on destroying North Sea shipping. A week after leaving base, the submarine was tracking a southward-bound convoy off Whitby, presumably waiting for an opportunity to attack, when the Armed Trawler *John Gilman* caught sight of its periscope and ran into ram. The U-boat realised the danger all too late and had insufficient time to dive deeply. The armed trawler grated over the U-boat, bending its periscope, and dropped two depth charges at point blank range. The force of the explosion crashed through the *John Gilman* throwing the crew around the deck but the underwater effect on the Germans must have been devastating. The *Miranda* III soon joined in the fray, dropping further depth charges, after which two more armed trawlers, *Viola* and *John Brooker*, steamed up to begin a hydrophone watch. A couple of hours later their vigil was rewarded: the two vessels glimpsed the shape of the U-boat moving just below the surface and both fired a couple of rounds as they closed, then depth-charged their prey as it slipped further downwards once more. A few minutes later, another Auxiliary Patrol trawler, this time the *Florio*, arrived on the scene and the stricken U-boat then surfaced amidst a black swell of oil and bubbles. The *John Gilman* and *Florio* dropped more depth charges as it sank for the final time. More oil and bubbles welled up but then subsided. The weight of evidence suggested the U-boat was finished, but no chances were taken: the seas were watched and later subjected to another depth charge attack, this time by the armed trawler *Sparrow*, at low water. An obstruction was later located, and four days later an Admiralty diving team confirmed the sinking.[64] The wreck site was later forgotten and only rediscovered by Scarborough divers in 1992.[65]

---

63 UB-107: uboat.net https://uboat.net/wwi/boats/index.html?boat=UB+107 (accessed 19 September 2018).

64 NA, ADM 137/2653, Anti-Submarine Division Monthly Report, 13 August 1918, and NA, ADM 137/805, 13 August 1918.

65 Robinson and Hart, *Viola*, 114.

Armed trawlers operating in the North Sea were involved in the destruction of one further U-boat before hostilities ended, and this action not only exemplified the effective way in which new equipment and weaponry was being deployed, but also illustrated the benefits that could accrue from joint action by different branches of the military. On 29 September 1918, a northward-bound convoy sailed at 1.29 p.m. from the Tyne under the escort of destroyers and armed trawlers. It was accompanied by the Airship R-29 from the East Fortune base, situated to the east of Edinburgh. Less than two hours later the airship observed oil rising to the surface a few miles off Beacon Point on the coast at Newton by the Sea in Northumberland, and dropped a 230 pound bomb, while radioing the senior officer of the escort force, the commander of HMS *Ouse*. The destroyer soon arrived on the scene but, after failing to locate the oil, requested the airship to drop a light to indicate the position of the suspected U-boat. This time, the airship let go both another bomb and a calcium flare. Now the *Ouse* bore down on the spot, firing off several depth charges. Soon afterwards the destroyer HMS *Star* joined in the fray, making a further depth charge attack to the westwards of the *Ouse*'s assault. The airship dropped a smoke bomb and further depth charges were fired by the *Star* after which more oil and large quantities of bubbles came to the surface.[66]

But the convoy was moving on, and first the *Ouse* and then the *Star* sailed northwards to renew their escort duties. By now six armed trawlers from the Tyne section, *Florio*, *Stronsay*, *Sparrow*, *Bombardier*, *Beatrice*, and *Viola*, which had been accompanying the convoy through their patrol section, joined in the hunt. Guided by the airship's activity, the first three of these vessels carried out another depth charge attack and buoyed the position while the latter three put their hydrophones over the side. Their watch was soon rewarded. At about 2.00 p.m., they picked up the loud sounds of a submarine emanating from the seas some two miles to the north-eastward. To the by now experienced ear of the hydrophone operator, the submarine sounded in a damaged condition. *Bombardier*, *Beatrice*, and *Viola* hauled in the hydrophones and proceeded line abreast over the suspect spot dropping twelve depth charges before resuming their hydrophone watch.[67]

The seas were quiet for some time, but at 3.10 the *Viola* heard strange noises in a north-north-east direction and proceeded to drop a further two depth charges, then started listening once more, while *Bombardier* searched the nearby surface waters, finding yet more oil bubbles. The hunt was maintained: more sounds were heard, and the destroyers HMS *Bonetta* and HMS *Tartar*, which had by now arrived on the scene, dropped additional depth charges.[68]

Still the pursuit continued. At 6.25, *Bombardier* picked up more sounds

---

66 NA, ADM 137/805, 29 September 1918.

67 Ibid.

68 Ibid.

still close. The enemy U-boat was clearly still fighting to escape in an eastwards direction, but the noises heard on the hydrophones suggested its machinery was operating with great difficulty. The machinery worked for only a few minutes at a time before seemingly jamming. *Bombardier* and section renewed their depth charge attack after which the undersea noises finally ceased.[69]

By now the armed trawlers had exhausted their depth charge reserves, so further supplies were sent out by drifter, while they maintained their hydrophone vigil throughout the night. At dawn, oil was seen rising freely to the surface. Although bad weather meant that an investigatory sweep of the area could not be carried out for several days, the kill was eventually confirmed. The Auxiliary Patrol report noted with satisfaction:

> The destroyers, airship and trawlers all worked together during the attack in a most able, efficient and harmonious manner which reflects greatest credit on all concerned. The airship especially distinguished herself by her smart lookout and efficient signalling.[70]

The Dover Barrage also demonstrated its effectiveness once more in late September 1918, when the UB-109 was blown up in the minefield when attempting to return to its Zeebrugge base. The vessel had left Bruges in late July and narrowly escaped destruction on its outward voyage through the Straits, having been sighted and forced to dive by three armed drifters. It had continued its voyage which took it as far south as the Azores and included the destruction of three merchant steamers. But as the submarine attempted to return to Flanders through the Straits on the surface during the early morning of 29 August, it was forced to dive amongst the mines owing to the presence of patrol vessels, and blew up. The eight survivors from the crew of 28 were picked up by the Auxiliary Patrol trawler *St Germain* which had narrowly escaped destruction on a German mine the previous year.[71]

The UB-109 was probably the last U-boat destroyed on the Dover Barrage, although at the time it was thought that UB-103 was sunk in a similar fashion after being driven by drifters onto the mines, but it now seems from recent research and investigation that this vessel had succumbed to mines laid much closer to the coast of Flanders on an outward voyage from Zeebrugge in August.[72]

By the autumn of 1918, most of the fishermen of the Auxiliary Patrol were experienced and well-trained veterans of the war at sea and their vessels much more effectively equipped than at any previous stage in the conflict. Those

---

[69] Ibid.

[70] Ibid.

[71] Chatterton, *Auxiliary Patrol*, 300–301.

[72] UB-103: uboat.net https://uboat.net/wwi/boats/index.html?boat=UB+103 (accessed 19 September 2018).

Photograph 11: Armed trawler *Lord Roberts c.*1915
Source: Charles Dinsdale

working in minesweeping vessels were more efficient in terms of the numbers of mines they removed than they had been in the first years of the war, while those involved in anti-submarine activities played a substantial and still often underestimated part of the stark maritime struggle. From the details outlined in this chapter, for example, fishing vessels on Admiralty service played a role in something in the region of 20 per cent of all allied U-boat sinkings during 1918. They were also working much more closely with other military units, both at sea and in the air, particularly through the convoy system, effectively to disperse and disrupt U-boat attacks on merchant fishing vessels. Moreover, the effectiveness of combined operations by all the forces working in the maritime theatre of war was responsible for the fall in the average monthly losses of merchant shipping during 1918, which never again approached the peaks witnessed in April and June 1917, despite the continued ferocity of the U-boat campaign (Photograph 11).

During September and early October, particularly after the submarine bases at Zeebrugge and Ostend were overrun by British land forces and the Germans abandoned their Adriatic base at Pola, U-boat activity became less marked but did not cease. Auxiliary Patrol forces everywhere were kept busy right up until the Armistice. In Malta, the end was marked somewhat laconically by Hull trawlerman William Oliver, skipper of the Grimsby minesweeping trawler *Kymric*. His log entry for that day reads:

Mon., Nov. 11th

11am. Received news re armistice. No orders, gave leave 2pm.[73]

This short entry, no doubt echoed in the logs of countless other Auxiliary Patrol armed fishing vessels, marked the end of more than four years of hostilities, not only for the thousands of fishermen in the RNR(T) and naval forces elsewhere but also for the inhabitants of every war-weary country that had been caught up in one way or another in this long and bitter conflict.

---

73 Signal Log, 1917–1918 kept by Skipper William Oliver while stationed in Malta. Preserved at the Blaydes Maritime Centre, Blaydes House, University of Hull.

Chapter Eight

The Aftermath

The First Months of Peace

In early 1919, the demobilisation of Auxiliary Patrol vessels and fishermen began. By this juncture, the bases from which the force had operated were already beginning to be run down, and many were completely closed during the following twelve months. The dismantling of this new navy was a major task involving thousands of people and vessels, and took some considerable time to organise and carry through to fruition.

Fishing vessels and crews eventually returned to their home ports from bases around the British Isles and beyond, and for those based overseas this was at times something of a trek. Skipper William Oliver, for example, who had been stationed in Malta for more than three years, during which time he had commanded the minesweeping trawlers *Marion* and then *Kymric*, was amongst the first to return from the Mediterranean, being given charge of the Grimsby Auxiliary Patrol trawler *Moravia* for the voyage back to England. On 12 December, just over a month after the Armistice, this minesweeper left Valetta Harbour for the final time, part of a flotilla of six trawlers and four drifters, their crews finally embarking on their homeward voyage. They were given what Oliver described as 'a wonderful send off, every ship in the harbour, blowing their whistles to which the homeward bound vessels responded in kind'.[1]

Progress across the Mediterranean was painfully slow: some of the drifters taking tows from the trawlers at various times in an attempt to maintain speed. It was not until nine days later, at noon on 21 December, that *Moravia* sighted the Rock of Gibraltar. After a short stay until the 24th, mainly for coaling and taking on stores, the flotilla sailed northwards. Despite this being the first December without war since 1913, the vessels were destined to spend Christmas Day at sea, season's greetings being sent to all ships from the flotilla's commanding officer aboard the trawler *Lemberg*. A day later the ships arrived in the River Tagus and anchored off Lisbon to snatch a brief sojourn, but a little more than 24 hours later they resumed their homeward voyage, finally sighting Lizard Point in heavy seas on the last day of 1918. Here the flotilla split: while

[1] Signal Log, 1917–1918 kept by Skipper William Oliver while stationed in Malta.

the drifters headed into Falmouth, the six trawlers proceeded in convoy along the Channel and by the morning of 2 January they were anchored in Portland Harbour.[2]

After five busy days in harbour, landing their hydrophone gear and other naval stores and equipment, Oliver took the *Moravia* back out to sea, voyaging onwards, along the English Channel then into the North Sea. At long last, early in the morning of 9 January 1919, only a little less than a month after leaving Malta, the trawler finally approached the Humber estuary, took on a pilot, and berthed in Grimsby Fish docks at 10.15 a.m. Those on board said their goodbyes, many taking trains for their home ports. For skipper, crew, and vessel alike, the war was effectively over; most would soon return to fishing.[3]

The story was much the same for many other Auxiliary Patrol vessels and their war-weary crews, who hailed from ports and harbours up and down the length and breadth of the British Isles. However, not all were fortunate enough to reach home in safety. The drifter *John Robert*, for example, disappeared while voyaging from Messina to Alexandria on 1 February 1919, presumed mined. No trace was found of either the ship or its 11 crew members, their only grave the sea.[4] Later the same month, the drifter *Philorth* foundered while voyaging from Syria to Malta, presumably on an early stage of its voyage home, though its crew seem to have survived.[5] A few other Auxiliary Patrol fishing vessels were also to be lost during this first year of peace, but mainly while on mine-clearance duties.[6]

Civilian fishermen who had been taken prisoner on the high seas also began to return. Twelve Grimsby skippers, including Alf Rawlings and George Blow. were amongst the earliest to get home, being repatriated from the Brandenburg Camp in Germany, by way of Copenhagen and Leith in late November. Many others had to wait much longer because of the sheer numbers involved, civilians and servicemen alike, but 18,000 men were reported to have already disembarked at the Riverside Quay in Hull by early December 1918.[7] Over 80,000 were to pass through the port before the repatriation process was completed.

Later, other fishermen came in through the Lincolnshire port of Boston, which had seen many crews taken when their vessels had been destroyed by the Germans in the first weeks of the war. In early January, Dutch vessels full of repatriated prisoners reached the Wash from Rotterdam, and anchored within six miles of Boston Docks. Although those on board could finally see their

---

2 Ibid.

3 Ibid.

4 Commonwealth War Graves Commission Records.

5 Aberdeen Maritime Museum, 'Aberdeen-Built Ships': http://www.aberdeenships.com/single.asp?words=Breivik&offset=2130&index=99730 (accessed 19 September 2018).

6 NA, ADM 186/604, 'History of British Minesweeping in the War', 85–86.

7 *Leeds Mercury*, 2 December 1918.

homeland again they had to spend one further night on board, but the following morning four tenders went down finally to bring them ashore. As they sailed towards the port on a clear and crisp January morning, the narrow banks of the river were packed with people. It seemed that almost everyone in the town had turned out to greet them. Cheers broke out when the tenders were sighted, those on board responding in kind, waving hands, crutches, and anything else: 'Are we downhearted?', came the cheer from one of the tenders. 'No', roared the masses on the river banks, and all began singing the popular wartime song 'It's a Long Way to Tipperary'. Emotions reportedly overflowed; more than 70 Boston fishermen had been held prisoner during the conflict and the townsfolk felt this homecoming deeply. Three of the tenders were decorated with bunting, but the fourth, full of the sick and injured, had a much more sombre appearance. There were soldiers and other sailors on board these vessels as well as fishermen. Special trains were waiting to move on the injured, while the rest were taken to the Fishermen's Institute where a good meal awaited. Any number of poignant reunions took place between local fishermen, their families, and friends, many having spent more than four years away from home. There was great joy amongst some, greeting family and friends they had not seen in some cases for more than four years, others meeting grandchildren and even children who had not been born when incarcerated. In other instances, the joy of getting back together was tinged with great sadness. During the incarceration of the Boston fisherman Louis Braime, taken back in 1914, two of his fishermen sons had been lost, one when minesweeping and the other while fishing. He was met by his wife and two daughters as another couple of sons were still away on war duties. Some expecting to be reunited with loved ones that day were disappointed: their relations had yet not returned, their reunions deferred for another day.[8] By evening, all those landed who were not from Boston had left by train on their journeys back home.

Although the war had ended, the casualty list continued to rise throughout the demobilisation process. Many seafarers who had survived the war died in the influenza epidemic by then sweeping across an exhausted world: Walter Kershaw, a Scarborough fisherman, who had been taken prisoner when the Hartlepool fishing vessel *Loch Ryan* had been captured by the Germans in September 1916, and was subsequently held at Dulmen Prisoner of War camp, returned home in December 1918, but within a week both his 24-year-old-wife Maria and three-year-old son Alfred had died after succumbing to the effects of the influenza epidemic.[9]

---

8 *Leeds Mercury*, 8 January 1918.

9 *Sheffield Daily Telegraph*, 7 December 1918 and *Aberdeen Evening Express*, 10 December 1918.

The Tragedy of the *Iolaire*

Fishermen and other mariners continued to lose their lives in war-related accidents, not least during the journey home. The worst such tragedy occurred in early hours of the morning of 1 January 1919, when HM Yacht *Iolaire* was wrecked in the Minch as it approached Stornoway Harbour. The vessel had taken on board many sailors and soldiers at the Kyle of Lochalsh, some possibly already demobbed, many others granted a period of leave. Most on board were sailors from the Western Isles who had covered the Christmas break for their English colleagues, preferring to take their own leave at New Year.[10] Hundreds of servicemen poured into the railhead, but the existing ferry service could not cope with the sheer numbers wishing to make the crossing, so soldiers and civilians were allocated places on the ferry *Sheila* while the *Iolaire* was drafted in to take aboard some 260 naval or RNR sailors. Many who embarked that night on the *Iolaire* were from the Isle of Lewis, most of the Harris men being told they would have to wait until the following day to make the crossing. The yacht sailed out of Kyle at 7.30 p.m. amidst driving wind and rain, which later turned into showers of sleet. At 1.55 a.m. on New Year's Day 1919, the vessel, nearing the end of its trip, struck the rocks known as the Beasts of Holm as it approached Stornoway Harbour. Early that morning, 201 men lost their lives, most of them Hebrideans; others were members of the *Iolaire*'s crew. Only 82 people survived. Forty of those who were saved reached safety behind a man who dropped over the side and made it to the shore with a line. He managed to crawl higher, above the waves, sitting with the line around his back while he pulled another couple of men ashore, then a heavier rope was attached and wrapped round a rock. The men on shore took turns to stabilise it, while 40 more men were pulled to the shore and safety.[11]

This tragedy was played out within yards of the shore – for so many so close to their native island. A lot of the bodies of men who had survived everything the Great War had thrown at them were now washed up by the sea on the nearby Sandwick shore, and hardly a family was left unaffected amongst the close-knit communities of Harris and Lewis. So many young men had died that it was said there was a shortage of coffins which had to be brought in from elsewhere. Many of those lost were RNR men and a substantial number were from the RNR(T) and had served in the Auxiliary Patrol. This is still the worst peacetime British maritime disaster since the loss of the *Titanic* and had a profound long-term effect on these island communities that had already lost more than 1,000 servicemen during the war.[12]

---

10 J. MacLeod, *When I Heard the Bell: The Loss of the Iolaire* (Edinburgh, 2012).

11 Stornoway Historical Society, '*Iolaire* Disaster': http://stornowayhistoricalsociety.org.uk/iolaire/ and 'Virtual Hebrides': https://www.virtualheb.co.uk/iolaire-disaster-western-isles/ (accessed 19 September 2018).

12 MacLeod, *When I Heard the Bell*.

Demobilisation

In home waters, Auxiliary Patrol convoy escort duties were mainly discontinued along the east coast after 12 November. Within days of the Armistice the task of lighting the East Coast War Channels with carbide lamps by a number of drifters was commenced, and soon completed – along much of the Yorkshire and Lincolnshire coasts, for example, by 23 November. Then, with the exception of vessels retained to maintain the lights until the channel was finally dismantled, the process of demobilising the Humber base drifter fleet began.[13]

Vessel demobilisation was at its height during the early months of 1919. As a rule, most of the easily removed fittings of war such as armaments, shells, depth charges, wireless equipment, and hydrophones were taken off before vessels finally left their bases, after which, all surplus coal stored in fish holds was removed so that vessels were ready for sending to demobilisation ports and centres set up to deal with the transfer back to civilian owners. The Admiralty ordered that vessels leaving their old base for the demobilisation centres should be supplied with a fortnight's worth of victualling provisions, or the same allowance in lieu, also deeming it ideal that craft arrive with the least possible amount of coal left on board.[14]

The necessity of surveying and reconditioning each vessel before it could be returned to its civilian owner meant that backlogs of trawlers and drifters built up at the demobilising ports. Gradually and steadily, however, the vessels were released and returned for fishing, their crews heading home with a railway travel voucher and a month's wages. The Auxiliary Patrol units around the coast were subsequently broken up and the bases run down. The Auxiliary Patrol Base at Tyne Dock, for example, was finally closed on 31 March 1919. The Lerwick Base lasted until mid-December 1919, being used that year as an intermediate base in British operations on the Murmansk coast.[15] By the end of 1919, the service was reduced to five Commands – Chatham, Portsmouth, Devonport, Rosyth, and Queenstown – although a few additional bases were left active to deal with the demobilisation of the last tranches of hired Auxiliary Patrol vessels.[16]

Mine Clearance Operations

Whilst many Auxiliary Patrol operations were being run down and the bases closed, there was one role which kept many crews and vessels active for much of 1919. This

---

13 NA, ADM 137/723, Auxiliary Patrol Humber, 16 November 1918 and 23 November 1918.

14 NA, ADM 1/85553/65, Demobilisation Supplement to Auxiliary Patrol List, 2–3.

15 *Aberdeen Journal*, 12 December 1919.

16 British Parliamentary Papers, Statement of the First Lord of the Admiralty, Explanatory of the Navy Estimates, 1919–1920, Cmd 451 (1919) xxxiii, 9–10.

was the removal of mines. During the war, 247,233 mines were laid in seas across the world, and by far the most (190,434) were laid in the North Sea, the largest number of which (71,126) were laid by British and American forces as part of the Northern Barrage. Although the British laid the largest number of mines, many in the later stages of the war as they constructed the great barrages, the Germans had sown no fewer than 25,782 mines in some 1,360 minefields or groups of mines in the seas around the British Isles and a further 17,854 mines in at least 60 separate minefields in other maritime theatres of war. Regardless of their origin, all these mines posed an undiscriminating threat to seafarers and their ships and needed to be neutralised before the high seas could be considered safe once more. Indeed, the danger, not only from minefields but also from drifting mines which had detached from their moorings, meant that the swept channels and minesweeping activities had to be maintained for some months after the return of peace.[17]

Even while the war was still raging, a programme of mine removal had been worked out in some detail, and on the conclusion of hostilities the Admiralty created a Mine Clearance Committee, inviting all maritime nations, apart from Germany and her former allies, to join up. Altogether, 26 countries signed up, and later, after the peace treaties had been ratified, the former enemy states were invited to join. Though mine clearance was under way soon after the Armistice in some areas, the holistic task of clearing mines from various areas was divided up amongst a number of these nations, and a large percentage of the work was allocated to the British. In order to carry this through effectively, a new Mine Clearance Service was created by the Admiralty in February 1919, and inaugurated on 1 April 1919.[18]

Before this date, during the first months of peace, efforts to remove mines within British spheres of influence put a great strain on the existing wartime minesweeping organisation. Discontent amongst ranks recruited for the hostilities only led to the Admiralty having to concede a liberal rise in pay to cover the risks being taken by minesweepers in the altered conditions, particularly as, in the case of fishermen at least, those working at their trade in this period were reported to be earning four times as much a week as ratings in the RNR(T).[19] As a temporary expedient, men were allowed to volunteer to serve on minesweepers for a minimum of three months, and although many came forward to work in this way, the rapid demobilisation of key personnel who did not wish to remain on military service sometimes meant that a vessel was unable to operate until they could be replaced. Some ships had a nearly full complement of crew drawing liberal wages, but were unable to participate in clearance operations because they were awaiting the appointment of a skilled rating.[20]

---

17 Dorling, *Swept Channels*, 337.

18 Ibid.

19 Boothby, *Spunyarn*, 235.

20 NA, ADM 186/604, 'History of British Minesweeping in the War', 85.

Many classes of seamen were eventually recruited to the new Mine Clearance Service. Even some soldiers volunteered.[21] And once again fishermen formed a substantial part of the backbone of this new organisation, which was under the general direction of Captain Lionel Preston, who had been associated with British minesweeping almost since its inception. Although initially ships and men employed in this post-war removal were from the minesweeping and patrol service, they too could demobilise if they wished, and their places were taken by the volunteers. Those involved were given special rates of pay and conditions of leave while being permitted to wear the Mine Clearance Badge, approved by the king, on their left sleeves. By June 1919, the service had reached its organisational zenith, employing 700 officers and 14,500 men in 55 different flotillas or groups of vessels. Home waters were divided into different areas under mine clearance officers who received their orders direct from the Admiralty. Abroad, Mine Clearance Officers overseas reported to Commanders in Chief. Altogether, the service's activities encompassed 22 different clearance areas around the coasts of the British Isles and many much further afield.[22]

One substantive exercise that was initially planned in December 1918 utilised 16 vessels together with skippers and fishermen drawn mainly from Stornoway, Lerwick, and Aberdeen, together with others from Hull and Lowestoft. This small fleet was under the charge of Commander Colin Inglis, DSO, RN, and swept mines from more than 3,000 square miles of seabed, while a special team was charged with destroying mines which had drifted ashore. Whilst most were fully engaged in sweeping operations, Inglis allowed one vessel, known as the 'duty drifter', to make a couple of fishing trips each week, and thus keep the rest of this sweeping flotilla supplied with fresh fish. These particular operations were in full swing by May 1919 and completed by the end of September.[23]

The British Mine Clearance Service worked in many and various overseas waters, including the Dutch coast, the Baltic, the Mediterranean, and the Black Sea. The removal of the substantial number of mines left in the Dardanelles after the initial swept channel had been created in the weeks following the Turkish Armistice in October 1918 was completed in July 1919. This proved a difficult operation, partly because of the currents, and partly because a number of the mines were laid in very shallow water, where conventional sweeping wires continually fouled on the bottom. One way of dealing with this problem was to use an observer in a kite balloon to locate each mine, which was then lassoed with a large wire loop. The crew members who did this became known as 'cowboys'. Having 'caught' the mine, the sweeper dragged it into slightly shallower water. Here the towing wire was released, and the mine bobbed to the surface where

21 Ibid., 85–86.
22 Ibid., 85.
23 Dorling, *Swept Channels*, 332–333.

rifle fire was used to finish it off. More than 3,000 mines were removed in the Dardanelles operations with the loss of four of the minesweepers. Nearly 50 crew members died in these operations.[24]

During the mine clearance activities in 1919, three minesweepers were lost in home waters, and at least two others in the Mediterranean. The numbers working in the service gradually reduced as people were paid off as areas were cleared. When the Mine Clearance Service had completed its work by the end of November 1919, British minesweepers had dealt with over 23,000 allied and more than 70 German mines since the armistice of the previous year.[25]

Post-War Losses of Fishermen and Fishing Vessels

Despite the efforts of the mine clearance forces in 1919, rogue sea mines continued to send both merchant ships and fishing vessels to the bottom in years to come. Fishing vessels were particularly vulnerable because they deployed and then hauled their extensive fishing gear across large swathes of the sea, and thus stood a considerable chance of fouling those mines which had escaped clearance activities. This was particularly a hazard for those fishing in North Sea waters, which had been so heavily mined during the war. It is impossible to ascertain just how many fishing vessels were lost in this way because although there were sometimes crew members who survived such encounters, or nearby ships that reported hearing or seeing a detonation in the area where a lost craft was known to be working, there were also other occasions when vessels and crews just disappeared without trace, and in some of these instances an unfortunate encounter with a stray sea mine may well have been behind the loss.

Three trawlers working out of Hull were definitely destroyed by mines in 1919. The first being the trawler *Sapphire*, which was blown up by a mine off Spurn Point on 20 February. The stricken trawler stayed afloat for 40 minutes after the explosion, which gave the crew plenty of time to take to their boat, eventually being picked up by the Hull trawler *Fidelia*.[26] Two of the crew had recently been demobilised from the Auxiliary Patrol.[27] Another recently demobilised fisherman, trimmer Joseph Raywell, was lost when his trawler *Scotland* went down, believed mined, off Flamborough Head in early March, and yet another RNR veteran, Bernard Bricknell, known as Bricky to family and friends, was making his first trip after demobilisation from minesweeping when the trawler *Durban* was lost a week or so afterwards.[28]

24 Ibid.

25 Ibid., 311.

26 *Hull Daily Mail*, 20 February 1919.

27 *Hull Daily Mail*, 21 February 1919.

28 *Hull Daily Mail*, 11 April 1919.

Another similar early peacetime loss was the Filey steam drifter *Emulator*, which struck a mine about 26 miles off Flamborough Head after leaving Scarborough harbour to go line fishing, and sank with all hands on 15 April 1919. At the time, *Emulator* was accompanying two other vessels that witnessed the explosion which blew the vessel to pieces. One of the crew, Richard Cammish, had also only recently returned from skippering a minesweeper in the Mediterranean.[29]

Sometimes fishing vessels were on hand to rescue mined sailors, as was the case with the Montrose drifter *Swift*, which rescued the five British crew of the German steamer *Hans Jost CT 9*, when the latter was blown up in late May 1919, some 54 miles north-eastwards of the Tyne, and landed them at Shields.[30]

In 1920, Scarborough lost three of its trawlers to mines, the first being the *Strathford*, which was blown up some 37 miles east-north-east of port on 27 February 1920, with the loss of all crew. Its skipper, Tom Normandale, had previously been skipper of the Scarborough trawler *Seal* when sunk by the U-57 while fishing in September 1916. The *Strathford*'s loss was followed on 9 May 1920 by the sinking of the *Taranaki*, which had its bows blown away by a mine caught in its nets when dropped onto the deck. One member of the crew was killed, but the rest were rescued by another Scarborough trawler. The third Scarborough trawler lost that year was the *Strathspey*, which was destroyed by a mine on 3 September, around 13 hours after leaving port. Its skipper, James Walker, had previously been on two Scarborough trawlers that were sunk by U-boats during the war.[31]

The losses continued. One of the vessels lost later was the Peterhead trawler *Adventure*, which succumbed to a mine caught in its nets about ten miles south-south-east of Buchan Ness in 1922. Luckily, the whole crew survived the explosion.[32] Some losses were perhaps avoidable, as was the case with the loss of Skipper Richard Whittleton and crew member James Roberts of the Hartlepool trawler *Gertude Cappleman*, in late October 1923. Their trawler had picked up a mine in its nets, but before throwing it back overboard the skipper had removed what was presumed to be the primer, which he took down into his cabin with the intention of keeping as a souvenir. The subsequent explosion killed both men and was put down to a lack of experience of mines by the skipper, who had been taken by the Germans off the Tyne in 1916 and held prisoner for the rest of the war.[33]

---

29 Scarborough Maritime Heritage Centre, 'Filey and the Great War – Minesweeping and Influenza': http://www.scarboroughsmaritimeheritage.org.uk/article.php?article=101 (accessed 19 September 2018).

30 *Dundee Courier*, 2 June 1919.

31 Scarborough Maritime Heritage Centre, 'Three Scarborough Trawlers Sunk by Mines in 1920': http://www.scarboroughsmaritimeheritage.org.uk/article.php?article=195 (accessed 19 September 2018).

32 *Yorkshire Evening Post*, 7 January 1922.

33 *Hartlepool Northern Daily Mail*, 2 November 1923.

By the end of June 1919, more than 1,000 English and Welsh fishing vessels had been demobilised. Many Scottish vessels were also released during the year, with just 131 being left in Admiralty service at the opening of 1920.[34] Many wartime restrictions on fishing had been quickly rescinded with the return of peace, but it took until the end of the 1919 to make some of the grounds safe for fishing. Fishing grounds around the British Isles were littered with wrecks and other debris of war, which created all sorts of problems, particularly because at first many of the sunken vessels were uncharted. Accordingly, fishing gear was lost or damaged, until fishermen became familiar with working on what had been until recently battlefields in the maritime war.[35]

The Impact of the Conflict on the Post-War Fishing Industry

Those who returned to fishing after demobilisation encountered a world much changed. Whilst methods of catching and processing fish had altered little during the war, the political and economic contexts in which the various sectors of their industry operated had been transformed in so many ways. The October 1917 Revolution which had brought the Bolsheviks to power had ultimately led to Russia severing many of its former international trading connections, while the markedly nationalistic approach adopted by many of the newly independent states which emerged in post-war Europe encouraged their governments to concentrate on nurturing their own trades and industries, often with the aid of subsidies and tariffs.

The new economic order was to have an almost immediate effect on the herring trade. Prior to the war, the British herring fisheries had been the most productive in the world, and much of their expansion had been based on the processing and export of salt-cured fish to markets across northern Europe. During the war, access to these markets was badly disrupted. Moreover, production had declined dramatically, not only because of enemy action but also because of the numbers of vessels taken out of fishing for Admiralty service and the pressing need for those vessels which remained fishing to try and maximise the fish supply for the home market.

For much of the inter-war years the Russian market was to remain almost entirely closed to the British herring industry, even though Norwegian fish exporters did re-establish and maintain a regular trade. The prospect of such channels of trade being revived receded even further after the Russians invested in the development of their own deep-sea fishing fleets during their first and second five-year plans. Trading with Germany was also plagued with

---

34 British Parliamentary Papers, Thirty-Eighth Annual Report of the Fishery Board for Scotland, Being for the Year 1919, Cmd 833 (HMSO, Edinburgh, 1920), vi.

35 Robinson, *Trawling*, 145.

problems. Soon after the return to peace the country's markets were flooded with Norwegian herring, and later in the 1920s trade was disrupted by dire currency problems.[36]

The full extent of overcapacity in the herring fisheries became all too evident when Government guarantees, which had maintained prices, were removed, and the period 1921 to 1924 proved especially difficult. The trade made a partial recovery in the later 1920s, but its longer-term export prospects were not helped by the determination of both Poland and Germany to improve their own capacity to catch fish. Despite some false dawns, the British herring fishery never came close to recovering its pre-war pre-eminence and prosperity. Output, so dependent on overseas exports, never again approached the record levels attained in 1913. The trade's underlying problems were aggravated by the effect of the Great Depression and the years between 1931 and 1933 proved especially difficult. One outcome of a subsequent Government investigation into the state of the entire British fishing industry was the creation of the Herring Industry Board, but in the very long term the traditional herring trade continued to decline and had virtually disappeared by the end of the 1950s.[37]

The trawling trade, the other leading sector of the pre-war British fishing industry, also encountered many problems in seeking to adjust to the markedly changed inter-war world. Initially, the trade boomed, even though it had to contend with internal labour disputes in some of the more significant trawling ports, as well as railway and coal strikes. North Sea fish stocks had recovered some of their vitality after more than four years of conservation through conflict. Despite problems caused by demobilisation and shortages of fishermen and vessels, 1919 landings were almost double the wartime average, and landings in the following year were some 14 per cent higher than had been attained in 1913.[38]

But the trawling trade was also soon in economic trouble. White fish prices declined rapidly during 1919, and then entered a long-term downward trend that lasted until the late 1930s. Although production costs fell, prices dropped even more rapidly. Trawling company profit margins were squeezed, which led to an almost constant pressure to cut labour costs in a trade where, unlike the herring and inshore sectors, the fishermen relied more heavily on the weekly wage than from a share of the proceeds.[39]

Such economic pressures had a profound effect. In England the port of Boston, already weakened by wartime sinkings, lost most of its deep-sea capability when its major trawling firm, the Boston Deep Sea fishing and Ice

[36] Jenny Sarrazin, 'War, Public Intervention and Structural Changes in the German Fishing Industry, 1914–1930', in Starkey and Tove Austbø, *Maritime Industries and Public Intervention*.

[37] Coull, *The Sea Fisheries of Scotland*, 220–221.

[38] Robinson, *Trawling*, 146.

[39] Ibid.

Company moved to the Humber in 1924.[40] Investment fell in many trawling ports. No newly built trawlers were added to the North Shields fleet between 1919 and 1929, and by 1930 almost one-third of the trawlers there were between 21 and 30 years of age. Although Granton still retained a substantial fleet of more than 80 trawlers in the early 1930s, trawling activity north of the border was ever more focused on Aberdeen, but even there the fleet was increasingly obsolescent. By 1934, approximately 60 per cent of the trawlers there were more than 20 years old.[41]

Many smaller and inshore coastal communities were caught by a vicious combination of falling prices and declining yields, and although productivity was improved for those who could afford vessels with motors, some found it difficult at times to cover fuel costs. In many smaller coastal communities, fishing was clearly a dying activity. By 1935, inshore communities contributed only 1 per cent of English and Welsh white fish landings, and around 15 per cent of Scottish landings. Such wider problems were reflected in an overall diminution in the size of the national labour forces. The numbers of English and Welsh fishermen declined from 42,555 in 1919, to 29,013 by 1936, and in Scotland they fell from 30,762 in 1921, to 21,480 in 1936.[42]

For much of the two decades that followed the Great War there was only one exception to this national catalogue of stagnation and decline. The Humber port of Hull continued to expand. Although it cut back on its middle water trawling activities, immediately after the war, first reducing the tonnage of vessels in its North Sea boxing fleet system by two-thirds in 1919, and then abandoning it altogether in 1936, the port's trawler owners invested heavily in the distant water sector. They continued to build modern, ever larger trawlers, opening and exploiting new grounds in the Arctic, and concentrating on the taking of cheaper and sometimes coarser varieties of white fish for the still nationally expanding fish and chip trade. This trend was eventually followed to some extent by the neighbouring port of Grimsby, which showed signs of revival from the mid-1930s after the opening of a new fish dock, and restarting a programme of investment in new trawlers.[43] The modern vessels in Hull and Grimsby's fleets were to be much in demand by the Admiralty once more when the Second World War broke out in 1939, and fishermen and fishing vessels from all over the country were again called up for service.

---

40 British Parliamentary Papers, Reports on English and Welsh Fisheries, 1924–1926.

41 R. K. Kelsall, H. Hamilton, F. A. Wells, and K. C. Edwards, 'The White Fish Industry', in M. P. Fogerty (ed.), *Further Studies in Industrial Organisation* (London, 1948), 108–115.

42 Robinson *Trawling*, 146.

43 Ibid., 156.

Fishermen in the Post-War Epoch

Many of those fishermen who had returned to fishing after demobilisation, men such as Joseph Watt VC of Fraserburgh, who, it will be recalled, was decorated for his action while serving on the Otranto Blockade, was one of many who returned to his home port and then lived and worked through the vicissitudes of these post-war decades. Fishing had always been an uncertain calling, and some, such as former Chief Skipper Charles Percy Allum, were still young enough to take up Admiralty service once more in 1939.[44] George Robinson from Hull, who spent part of the war minesweeping on HMS *Bourne* in waters off Harwich, was finally demobilised in June 1919 and returned to working as mate in the Hull North Sea boxing fleets, until the last of these operations ceased in 1936. Perhaps too old by then for fishing in the distant water sector, he joined the tanker *Strafford* as Ordinary Seaman, and spent three years running supplies into government forces fighting General Franco in the Spanish Civil War. The Second World War found him on Atlantic convoys until his health gave out in 1943. He died the following year aged 60. Many others were also too old for Second World War service or never made it that far: William Oliver, the Hull trawler skipper who had served on minesweepers off Malta for much of the conflict, spent most of the inter-war period trawling in Arctic waters, but had left the sea by the outbreak of the Second World War. Later, in 1945, he became secretary of the Hull Trawler Officers' Guild.[45]

George Ernest Stubbs, decorated for his role as skipper of the *Coreopsis* in the sinking of the U-85, became licensee of the Dock Tavern in Gorleston in 1922,[46] but evidently did not completely give up his former calling, his death at sea aged 47, while skipper of the Yarmouth drifter *Cheerio Lads*, being reported at Lerwick in June 1934.[47] His commander, Lieutenant Peat, not a fisherman, but a prominent player in the Auxiliary Patrol, became assistant harbour master at Shields in 1923, but died aged only 45 in 1936. Skipper George William Tharratt, decorated for minesweeping, also went back to fishing on trawlers, working out of the Humber after serving in the Mine Clearance Service, but he died of a stricture in 1928, leaving a partner and five young children. Just 11 years after receiving the DSC he was buried in a pauper's grave, not the first and certainly not the last fisherman to be so.[48]

Fred Simmonds, born in 1898, from the village of Chickerell by the Dorset Coast, had signed up in February 1917, and spent a great deal of his service in the Mediterranean. He had had a spell in hospital in Malta suffering from

44 Robinson and Hart, *Viola*, 128.
45 Robinson, *Far Horizons*, 56–58.
46 *Yarmouth Independent*, 7 January 1922.
47 *Dundee Courier*, 18 June 1934.
48 Robinson and Hart, *Viola*, 122.

malaria, then spent much of the spring and summer of 1918 serving in Taranto on HMT *Checksin*, a German trawler which had been captured by the Admiralty in 1915. He was demobilised in February 1919. Albert Parker, from the same Dorset village, had married Alison Constable by special licence in the parish church in July 1918, having been called up for military service. He enrolled in the RNR just three days later. He too served at Taranto, spending some time on the Admiralty trawler *Charles Chappell*, as well at Mudros and other places in the Mediterranean. Albert, or Bert as he was known, was finally demobilised in August 1919, possibly having spent time on mine clearance work after the end of hostilities. Both men returned to their home village and lived and worked in the area for the rest of their lives, Bert making his subsequent living by keeping a small herd of cows and delivering milk.[49]

Others moved in different directions. Magnus Flaws, a fisherman from Dunrossness, Shetland who served at one time as Mate on the armed trawler *Viola*, later spent some time after the Armistice as part of the forces based at Constantinople while the mine clearance operations were in full swing.[50] He returned for a short while to Shetland after demobilisation but, like many from the Scottish Isles, joined the merchant navy, and by 1930 was working on vessels trading along the Australian coast. When the slump hit hard he bought a £33 ticket to London and then returned to Shetland where he married and went back to fishing, working as a driver (engineer) in the herring fishing industry for the rest of his seagoing career.[51]

Few were unaffected by their war service in one way or another. Like many other people caught up in this first total war, the patterns and sometimes the expected course of their future lives were altered forever. The last vestiges of the Auxiliary Patrol, the service in which they served, was fragmented, and the RNR(T) was abolished in 1921 as a separate entity of the RNR. Substantial numbers of fishermen, however, continued as naval reservists and many of the next generation and their fishing vessels were to be called up to form part of the new Royal Naval Patrol Service, created after the Second World War broke out in 1939.

---

49 I am grateful to Joy Simmonds for the biographical information collected on her grandfathers, Fred Simmonds and Bert Parker.

50 NA, BT 377/7667, Magnus Flaws RNR Service Record.

51 Biographical details supplied by the Flaws family, Sumburgh, Shetland.

Epilogue: Contribution and Cost

In 2011, a company specialising in organising London events planned to erect a large marquee and hold a series of banquets and parties, presumably for people working in the City, on the run up to Christmas, in Trinity Square Gardens, Tower Hill, London. On one side of these gardens lies Sir Edwin Lutyens's memorial to 35,000 fishermen and merchant seamen killed in the Great War, who have no grave but the sea. Behind it is the sunken garden, which records the names of those seafarers who died during the Second World War on Atlantic and Antarctic Convoys, and nearby is a memorial to sailors lost in the more recent Falklands conflict. The idea of using such a place as a site for drinking and entertainment created a degree of controversy and indignation in the press and amongst some sections of the public, and was eventually abandoned. That it was ever mooted tells us something of the way in which the contribution of fishermen and indeed many other non-mainstream seafarers who served in one way or other outside of the mainstream Senior Service has been to such a great extent forgotten. One cannot imagine such plans would ever have even been mooted for somewhere like the Cenotaph in Whitehall, the Commonwealth war grave cemeteries in France, or many other national war memorials.

This collective amnesia when it comes to the role of the fishermen and other groups was not initially evident. Immediately after the war and in the first years which followed, the crucial role that the fishermen played was often acknowledged by key figures in the country's political and military establishment. Moreover, a small number of books highlighting their activities in the Great War were published, which told much of the scale and scope of their contribution. The new navy known as the Auxiliary Patrol had been created very quickly: in no more than a few months at the beginning of the war. It also was dispersed and disbanded very rapidly after the return of peace. Few fishermen of the day later put pen to paper to reflect on their wartime experiences. In later decades, while an ocean of academic print has been produced about dreadnoughts, Jutland and the like, often by mainstream Royal Navy historians, all too often the role of the fishermen and fishing industry, as well as indeed many other groups of seafarers, has been either ignored, underplayed, misunderstood, or at the very best understated. This has continued in many ways during the commemorations to mark the centenary of the conflict. Indeed, it has been possible to watch many media presentations or visit many of the national exhibitions devoted to the centenary of the Great War and never come away with any real idea of the scale and scope of the engagement of fishermen and coastal communities from around the British Isles on the maritime front line.

For an island nation, so dependent in many ways for so long on the sea, this may seem to some extent surprising. However, it has not been the purpose of this book to analyse the reasons for this collective and continuing amnesia, but to attempt to understand the reasons for – and assemble some reminders about the scale and scope of – the fishing industry's engagement, and it is perhaps apt at this point to outline some key aspects of this.

Altogether, 1,467 steam trawlers and 1,502 drifters were hired or requisitioned by the Admiralty for war service.[1] They were drawn from ports along almost every part of the British Isles. In 1917, a further 300 fishing vessels were also taken up for the fishery reserve, the idea here being to provide more armed protection for vessels still fishing.[2]

As we have seen, when fishing vessels were taken up by the Admiralty their extant crews often went with them, although many of these fishermen went on to serve on other former fishing vessels and craft deployed by the Auxiliary Patrol and related forces. In 1920, the English Fisheries Board claimed that no British occupation contributed a greater portion of its workforce to the war effort.[3] In 1913, there were just over 100,000 full- or part-time fishermen in Britain. Official figures show that at one time or another 49 per cent of fishermen were on Admiralty service during the conflict, while a significant number of others had volunteered for army service in the early months of the conflict.[4] Most of those who continued fishing pursued their calling off the British coasts or further afield on what was for all intents and purposes a maritime front line.

Although their role on the front line of minesweeping duties is often recognised, it has not been written about to any great extent. Criticisms have sometimes been levelled at the effectiveness of fishing vessels and similar craft of the Auxiliary Patrol in terms of a poor ratio of U-boats destroyed in terms of the numbers of vessels deployed.[5] However, fishing vessels on Admiralty service probably participated more substantially in incidents of U-boat destruction than

---

1 Robb Robinson, 'A Forgotten Navy: Fish, Fishermen, Fishing Vessels and the Great War at Sea', *Journal for Maritime Research* 19.1 (2017): 47–61.

2 See Chapter 7.

3 British Parliamentary Papers, Fisheries England and Wales: Fisheries in the Great War: Being the Report on Sea Fisheries for the Years 1915, 1916, 1917 and 1918 of the Board of Agriculture and Fisheries, vii.

4 Robinson, *Trawling*, 136–137.

5 See, for example, *Marder, From Dreadnought to Scapa Flow*, 2:357 and Julian Thompson, *The Imperial War Museum Book of the War at Sea, 1914–1918* (London, 2005), 327.

has sometimes been acknowledged, and in any case basing their effectiveness on the number of U-boat sinkings is a rather narrow criterion on which to judge this dimension of the Great War at sea. The Auxiliary Patrol in which they were deployed made extensive use of later incoming technology such as hydrophones and depth charges. It played a fundamental part in making coastal convoys so effective after they were introduced and worked closely with airship and aircraft patrols to disrupt and deter U-boat attacks, often forcing them to take evasive action when trying to make attacks on shipping. In the later stages of the war not only were more U-boats sunk in attacks, but many others must have been damaged to the extent that they needed to curtail operations and return to base for extensive repairs. Above all, the war channels around the British Isles were kept open. Britain and the many theatres of war in which its forces operated were kept supplied with vital consignments of food, raw materials, and military equipment. In short, the country's ability to continue the war was not degraded.

Whilst substantial numbers of fishermen in some regions, particularly in places such as the Western isles, were already in the RNR, many who joined up from August 1914 entered the Royal Naval Reserve (Trawler Section) whose numbers grew exponentially. Prior to the outbreak of war there were just 140 vessels and 1,278 skippers and fishermen in the Trawler Reserve, but, as we have seen, by the Armistice there were not only more than 3,000 fishing vessels taken into service but more than 39,000 fishermen in what was called the Auxiliary Patrol.[6]

As Figures 11, 12, and Table 3 illustrate, most fishing vessels were lost during the conflict in the North Sea, but the figures also illustrate something of the spread of vessel activity. Whilst submarines were the main source of losses amongst working fishing vessels, mines accounted for a significant number of Admiralty craft. As we have seen in the preceding chapters, fishing vessels not only played an essential role in terms of minesweeping and keeping the sea lanes – what became known as the war channels – clear around the coasts of southern and eastern Britain and beyond, but they often played key roles in important theatres of the maritime conflict far removed from home waters.

During more than four years of conflict, 672 fishing vessels and 416 civilian fishermen were lost while continuing to fish and making an important contribution to the hard-pressed nation's food supply. Most ports lost a lot of vessels: Fleetwood, for example, lost 60 vessels out of a pre-war fleet of 150 trawlers. A total of 3,338 skippers and ratings were also lost while serving with the Auxiliary Patrol. But these figures understate markedly the total losses because many other fishermen had, of course, joined the traditional Royal Naval Reserve, and it is currently difficult to identify all the RNR fishermen from the existing casualty data. By the end of the war there were no more than 14,000 fishermen left fishing from English and Welsh fishing stations and of these 8,000 were above military age and 400 below. Despite this marked reduction in workforce and vessels the

6 Robinson, 'A Forgotten Navy', 47 and 59.

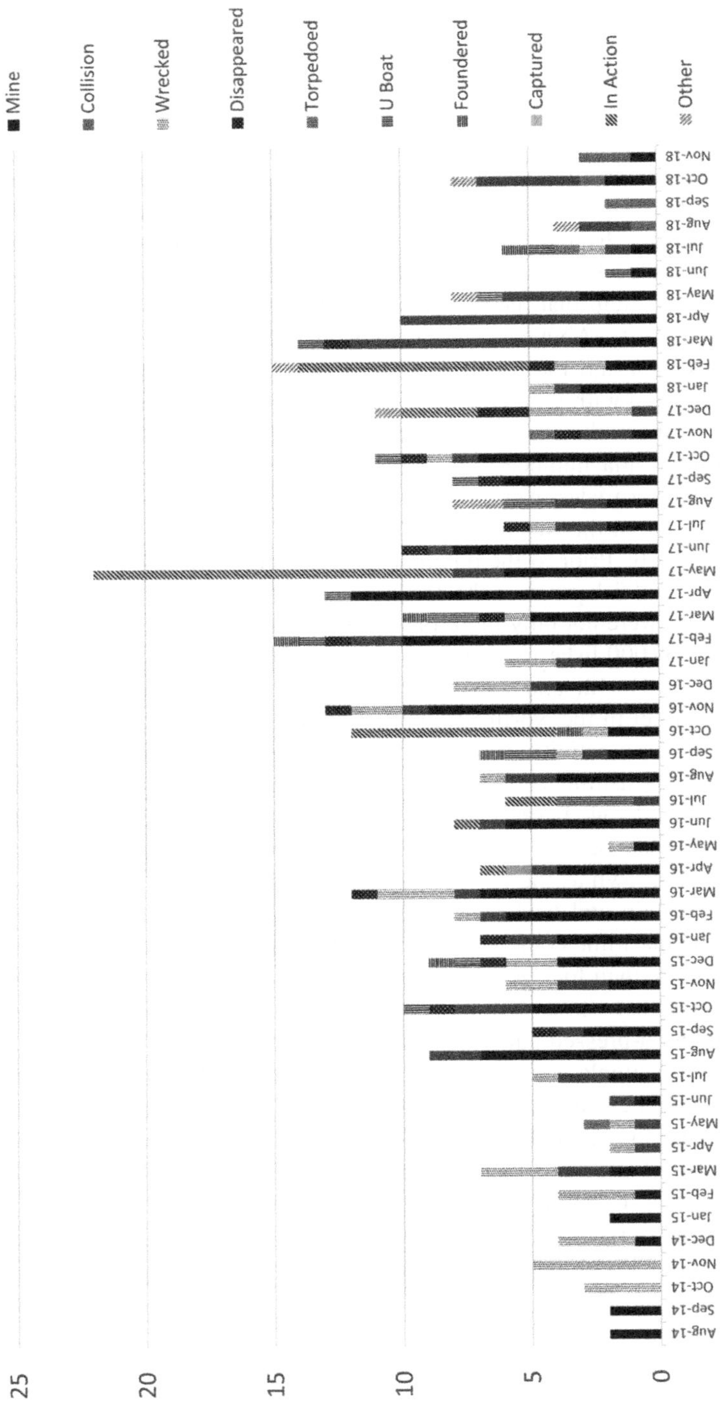

Figure 11: Losses of fishing vessels on Admiralty service, August 1914–November 1918
Source: HMSO, *British Vessels Lost at Sea, 1914–1918* (London, 1919).

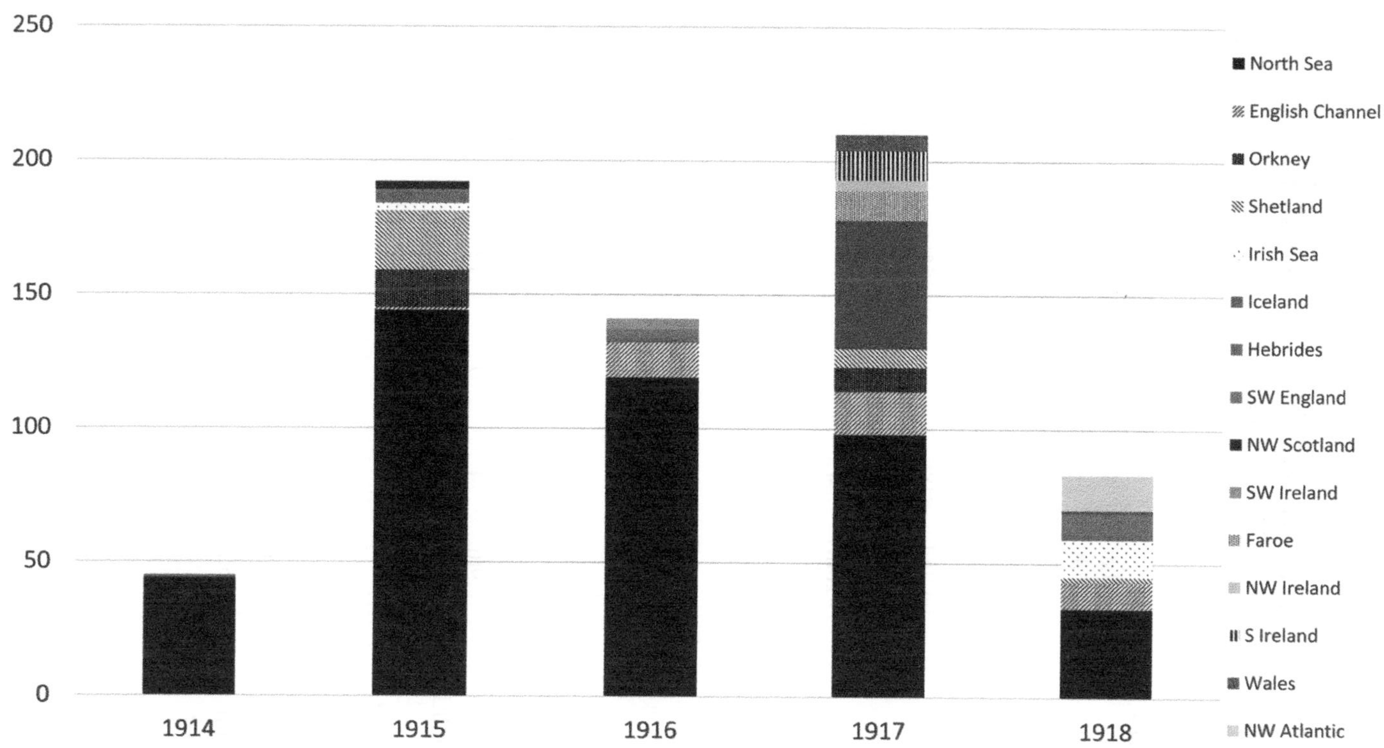

Figure 12: Locations of fishing vessels destroyed by enemy action, 1914–1918
Source: HMSO, *British Vessels Lost at Sea, 1914–1918* (London, 1919).

Table 3:
Disposition of minesweeping forces, November 1918 (home waters)

Auxiliary Patrol Area	Fast sweeping flotillas		Hired paddlers	Trawlers	Drifters	Dance class
1. Stornoway	–	–	–	6	–	–
2. Oban	13th	7				
3. Swarbacks Minn			–	2		–
Lerwick	21st	4		6	4	
Kirkwall	4th 5th			5	17	
Scapa				24		
4. Cromarty	–	–	–	10	4	–
5. Peterhead	–	–	–	23	17	–
6. Granton	1st 2nd	18	–	34	–	–
	3rd	14				
	16th	8				
	20th	4				
8. Tyne			8	31		
9. Grimsby	7th	8	4	22	6	
10. Lowestoft				42	8	
Harwich	6th	4	6	19	16	
Nore			2	24	6	
11. Ramsgate				8		
Dover	10th	5	7	35		
Dunkirk						10
12. Newhaven				5	9	–
Portsmouth	9th	4	7	12		
Yarmouth IoW	22nd	4			8	
Havre				14		
Cherbourg				6		
13. Portland	17th	6	–	5	–	–
13a. Torquay			–		8	–
Devonport	18th	4		12		
14. Falmouth	15th	4	–	6		–
Penzance				6	6	
15. Milford			–	12		–
Bristol Channel	19th	4		5		
16. Kingston				6		–
Liverpool	11th	4	5	9	5	

Auxiliary Patrol Area	Fast sweeping flotillas		Hired paddlers	Trawlers	Drifters	Dance class
17. Lough Larne			8			–
18. Lough Swilly			–	12	6	–
19. Clyde	12th	4	–	7	12	–
20. Berehaven			–	4		–
21. Queenstown			–			–
Waterford	8th	4		3	3	
22. Holyhead	–		–	2		
Totals		110	47	417	135	10

Source: National Archives, ADM 186/604.

British fishing industry's landings in 1918 remained at 31 per cent of pre-war output, while the German output had slumped to no more than 9 per cent over the same period.[7]

Many of those lost, as we have noted, have no grave but the sea. Their names are marked in places such as the Tower Hill Memorial and with many other servicemen on war memorials throughout the British Isles. But, unless they died in sea areas far removed from home waters, many of those who have war graves are not usually buried in war cemeteries such as those found behind the Western Front in France and Belgium. Many are to be found in quiet corners of cemeteries and churchyards across the length and breadth of the British Isles, most notably in coastal ports, towns, and villages, often quite close to where their bodies were brought ashore, an indication of the scale and scope of their contribution to the Great War at sea.

[7] Robinson, 'War, Public Intervention', 126 and Sarrazin, 'War, Public Intervention', 147–149.

Select Bibliography

Primary Sources

Admiralty Publications: Printed Instructions, Monographs and Reports 1827–1970 (ADM 186), National Archives, Kew

Admiralty, and Ministry of Defence, Navy Department: Correspondence and Papers, Demobilisation Supplement to Auxiliary Patrol List (ADM 1/85553/65), National Archives, Kew

Admiralty, and Ministry of Defence, Navy Department: Ships' Logs, 1799–2003 (ADM 53), National Archives, Kew

Admiralty, Historical Section: Papers Used for the Official History of the First World War (ADM 137), National Archives, Kew

Diary of G. Robinson, 2nd Hand, HMT *Edward VII* from the day of December 10th when joined up for sweeping enemy mines in the North Sea in the year 1915 & after (original in possession of author)

North-Eastern District Sea Fisheries Committee Reports, 1914–1922, The Treasure House, Beverley

Royal Naval Reserve Service Records, Catalogue Reference BT/377, National Archives, Kew

Signal Log, 1917–1918 kept by Skipper William Oliver while stationed in Malta, Blaydes Maritime Centre, the University of Hull

Official Reports

British Parliamentary Papers, Fisheries England and Wales: Board of Agriculture and Fisheries, Annual Reports on Sea Fisheries for the Years 1913–1926

British Parliamentary Papers, Reports of the Fishery Board for Scotland, 1913–1922

British Parliamentary Papers, Statement of the First Lord of the Admiralty, Explanatory of the Navy Estimates, 1919–1920, Cmd 451 (1919)

Hansard

Newspapers

Aberdeen Press and Journal
Aberdeen Weekly Journal
Daily Record
Dundee Courier
Grimsby Evening Telegraph
Hartlepool Northern Daily Mail
Hull Daily Mail
Leeds Mercury
Liverpool Daily Post
London Daily News
London Gazette
Manchester Courier and Lancashire General Advertiser
Pall Mall Gazette
Sheffield Daily Telegraph
The Times
Thetford & Watton Times and People's Weekly Journal
Yarmouth Independent
Yorkshire Post and Leeds Intelligencer
Yorkshire Evening Post

Books and Articles

Sir R. Bacon, *The Dover Patrol*, vol. 1 (London, 1919).
—— *The Dover Patrol*, vol. 2 (London, 1919).
Tim Benbow, *Naval Warfare, 1914–1918: From Coronel to the Atlantic and Zeebrugge* (London, 2008).
H. B. Boothby, *Spunyarn* (London, 1929).
Edward Keble Chatterton, *Q Ships and Their Story* (London, 1922).
—— *The Auxiliary Patrol* (London, 1923).
—— *Danger Zone* (London, 1934).
L. Cope Cornford, *The Paravane Adventure* (London, 1919).
Julian Corbett, *History of the Great War: Based on Official Documents, Naval Operations*, vol. 1 (London, 1920).
—— *History of the Great War: Based on Official Documents, Naval Operations*, vol. 2 (London, 1921).
—— *History of the Great War: Based on Official Documents, Naval Operations*, vol. 3 (London, 1923).
James R. Coull, *World Fishery Resources* (London, 1993).
—— *The Sea Fisheries of Scotland: A Historical Geography* (Edinburgh, 1996).
R. Crompton-Hart, *Submarines at War, 1914–1918* (London, 1970).
Taprell Dorling, *Swept Channels* (London, 1935).
Michael Epkenhans, Jörg Hillmann, and Frank Nägler (eds), *Jutland: World War I's Greatest Naval Battle* (Lexington, Ky, 2015).

S. J. Eyre, *Skippers of the Royal Naval Reserve in the First World War* (Leamington Spa, 2014).

Antony Firth, *East Coast War Channels in First and Second World Wars*, A Report for English Heritage (Tisbury, 2014).

Norman Friedman, *Fighting the Great War at Sea: Strategy, Tactic and Technology* (Annapolis, Md, 2014).

P. Gariepy, *Gardens of Hell: Battles of the Gallipoli Campaign* (Lincoln, Nebr., 2014).

R. H. Gibson and M. Prendergast, *The German Submarine War* (London, 1931).

Alec Gill, *Lost Trawlers of Hull: Nine Hundred Losses between 1835 and 1987* (Beverley, 1987).

R. M. Grant, *U-Boats Destroyed* (London, 1964).

Paul Halpern, *The Battle of the Otranto Straits* (Bloomington, Ind., 2004).

HMSO, *British Vessels Lost at Sea, 1914–1918* (London, 1919).

Richard Hough, *Dreadnought: A History of the Modern Battleship* (Penzance, 2003).

Sir Archibald Hurd, *History of the Great War: The Merchant Navy*, vol. 1, *1914 to Spring 1915* (New York, 1921).

—— *History of the Great War: The Merchant Navy*, vol. 2 (New York, 1921).

—— *History of the Great War: The Merchant Navy*, vol. 3 (London, 1924).

N. Hyman, 'Inshore Fisheries in Wartime Devon', in H. French, *Food, Farming and Fishing in Devon during the First World War* (Exeter, 2017).

Admiral Viscount J. Jellicoe, *The Grand Fleet, 1914–1916* (London, 1919).

David T. Jones, 'Scottish Fisheries during the War', in David T. Jones, Joseph F. Duncan, H. M. Conacher, and W. R. Scott (eds), *Rural Scotland during the War* (London, 1926).

R. K. Kelsall, H. Hamilton, F. A. Wells, and K. C. Edwards, 'The White Fish Industry', in M. P. Fogerty (ed.), *Further Studies in Industrial Organisation* (London, 1948).

E. F. Knight, *The Harwich Naval Forces: Their Part in the Great War* (London, n.d.).

Rotem Kowner, *Historical Dictionary of the Russo-Japanese War*, 2nd edn (London, 2017).

—— 'The Impact of the War on Naval Warfare', in Rotem Kowner (ed.), *The Impact of the Russo-Japanese War* (London, 2007).

Andrew Lambert, 'Tradition, Technology and Transformation', in Robert J. Blyth, Andrew Lambert, and Jan Rugger (eds), *Dreadnought and the Edwardian Age* (Oxford, 2011).

Nicholas A. Lambert, *Sir John Fisher's Naval Revolution* (Columbia, SC, 2002).

Howard S. Levie, *Mine Warfare at Sea* (London, 1992).

Samantha Little, 'The Brixham Fishing Fleet in the Great War', in H. French, *Food, Farming and Fishing in Devon during the First World War* (Exeter, 2017).

Innes McCartney, 'Paying the Prize for the German Submarine War', *Mariner's Mirror* 104.1 (2018).

S. McGreal, *Liverpool in the Great War* (Barnsley, 2014).

F. M. McKee, 'An Explosive Story: The Rise and Fall of the Common Depth Charge', *Northern Mariner* 3.1 (1993): 45–58.

J. MacLeod, *When I Heard the Bell: The Loss of the Iolaire* (Edinburgh, 2009).

Arthur Marder, *From Dreadnought to Scapa Flow*, vol. 2, *To the Eve of Jutland 1914–1916* (Oxford, 1969).

—— *From Dreadnought to Scapa Flow*, vol. 4, *1917, Year of Crisis* (Oxford, 1969).

G. H. P. Mulhauser, *Small Craft* (London, 1920).

Henry Newbolt, *History of the Great War*, vol. 4, *Naval Operations* (London, 1928).

James Nicolson, *Food from the Sea* (London, 1979).

Roger Parkin, *Dreadnought: The Ship that Changed the World* (London, 2015).

Tony Pawlyn, 'Fisheries of the West Country', in David J. Starkey, Chris Reid, and Neil Ashcroft (eds), *England's Sea Fisheries: The Commercial Sea Fisheries of England and Wales since 1300* (London, 2000).

G. C. Peden, *Arms, Economics and British Strategy: from Dreadnoughts to Hydrogen Bombs* (Cambridge, 2007).

David Ramsey, *The Lusitania Saga and Myth* (Barnsley, 2015).

Chris Reid, 'From Boom to Bust: The Herring Industry in the Twentieth Century', in David J. Starkey, Chris Reid, and Neil Ashcroft (eds), *England's Sea Fisheries: The Commercial Sea Fisheries of England and Wales since 1300* (London, 2000).

Jurgen Rhower, 'Submarine Warfare: Central Powers', in Spencer Tucker (ed.), *The European Powers in the First World War: An Encyclopaedia* (London, 1999).

George F. Ritchie, *The Real Price of Fish: Aberdeen Steam Trawler Losses, 1887–1961* (Beverley, 2001).

Robb Robinson, *Trawling: The Rise and Fall of the British Trawl Fishery* (Exeter, 1996).

—— 'The British Distant Water Trawling Industry 1880–1939', in David J. Starkey and Alan Jameson (eds), *Exploiting the Sea: Aspects of Britain's Maritime Economy since 1870* (Exeter, 1998).

—— *Far Horizons: From Hull to the Ends of the Earth* (Hull, 2010).

—— 'The North Sea Littoral and the British Isles', in David J. Starkey and Ingo Heidbrink (eds), *A History of the North Atlantic Fisheries*, vol. 2, *From the 1850s to the Early Twenty-First Century* (Bremen, 2012).

—— 'Alward, George Lowe (1842–1933), trawler owner', *Oxford Dictionary of National Biography* (Oxford, 2016).

Robb Robinson and David J. Starkey, 'The Sea Fisheries of the British Isles, 1376–1976: A Preliminary Survey', in Poul Holm, David J. Starkey, and Jón T. Thór (eds), *The North Atlantic Fisheries, 1100–1976: National Perspectives on a Common Resource* (Ejsberg, 1996).

Robb Robinson and I. Hart, *Viola: The Life and Times of a Hull Steam Trawler* (London, 2014).

Nicholas A. M. Rodger, 'Anglo-German Naval Rivalry 1860–1914', in Michael Epkenhans, Jörg Hillmann, and Frank Nägler (eds), *Jutland: World War I's Greatest Naval Battle* (Lexington, Ky, 2015).

Jenny Sarrazin, 'War, Public Intervention and Structural Changes in the German Fishing Industry, 1914–1930', in David J. Starkey and A. Tove Austbø (eds), *Maritime Industries and Public Intervention, The Fourth North Sea History Conference* (Stavanger, 2002).

Lawrence Sondhaus, *Naval Warfare, 1815–1914* (London, 2001).
M. Stille, *The Imperial Japanese Navy of the Russo-Japanese War* (Oxford, 2016).
Pat Sweeney, *Liffey Ships and Shipbuilding* (Cork, 2010).
Julian Thompson, *The War at Sea, 1914–1918* (London, 2005).
Gerald Toghill, *Royal Navy Trawlers, part 1: Admiralty Trawlers* (Liskeard, 2003).
—— *Royal Navy Trawlers, part 2: Requisitioned Trawlers* (Liskeard, 2004).
Spencer Tucker (ed.), *A Global Chronology of Conflict: From the Ancient World to the Modern Middle East* (Oxford, 2010).
Spencer Tucker and T. J. Weiler, 'The Arctic Theater', in Spencer Tucker and Priscilla Mary Roberts, *World War I: The Definitive Encyclopedia and Document Collection* (Santa Barbara, Calif., 2015).
Martin Wilcox, 'Concentration or Disintegration? Vessel Ownership, Fish Wholesaling and Processing in the British Trawl Fishery, 1850–1939', in David J. Starkey and James E. Candow (eds), *The North Atlantic Fisheries: Supply, Marketing and Consumption*, 1560–1990 (Hull, 2006), 56–57.
H. W. Wilson, *Hush or the Hydrophone Service* (London, 1920),

Acknowledgements

It is not possible to mention everyone who has encouraged the writing of this book or provided advice and much welcome support along the way. First and foremost, I would like to thank the many people and families whose ancestors were involved in various aspects of the maritime war as well as those other individuals who have also been so obliging in providing access to photographs, relevant documents and recollections or who have supplied such unstinted material help and advice during the processes of research and writing. These include: Jim Anderson, Skipper Peter Craven, Alan Curtis, Charles Dinsdale, Skipper Trevor Doyle, Paul Escreet, Antony Firth, Andrew Flaws, Jon Grobler, Pat Hindshaw, Alan Hopper, Phil Johnson, Skipper Ken Knox, Kim Lever, Peter Mason, James Morley, Professor Hugh Murphy, John Nicholls, Carl Racey, Bruce Robinson, Joy Simmonds, Captain John Simpson, Brian Smith, Professor David J. Starkey, Ian Tait, Michael Teare, Eric Tharratt, Pat Tharratt, Mark Vesey, Billy Wells, Richard Weymss, George Westwood, and Dr Martin Wilcox.

My thanks are also due to staff at the British Library, the Brynmor Jones Library at the University of Hull, the Hull History Centre, the National Archives, the Scarborough Maritime Heritage Centre, the Scottish Fisheries Museum, the Shetland Museum and Archives, and the Treasure House, Beverley.

I also owe Peter Chapman of the East Yorkshire Family History Society a particularly deep debt of gratitude for his invaluable and unstinting help in terms of genealogical issues and his supporting background research relating to tracking down the full background of so many individuals and incidents encompassed by this study.

My deepest thanks, however, are due to my wife, Wendy, whose endless patience and close and constant encouragement has enabled me to take on these endeavours and see them through to fruition.

Index